3D Point Cloud Analysis

Shan Liu • Min Zhang • Pranav Kadam
C.-C. Jay Kuo

3D Point Cloud Analysis

Traditional, Deep Learning, and Explainable Machine Learning Methods

Shan Liu
Tencent Media Lab
Palo Alto, CA, USA

Min Zhang
University of Southern California
Los Angeles, CA, USA

Pranav Kadam
University of Southern California
Los Angeles, CA, USA

C.-C. Jay Kuo
University of Southern California
Los Angeles, CA, USA

ISBN 978-3-030-89182-4 ISBN 978-3-030-89180-0 (eBook)
https://doi.org/10.1007/978-3-030-89180-0

This Springer imprint is published by the registered company Springer Nature Switzerland AG
The registered company address is: Gewerbestrasse 11, 6330 Cham, Switzerland

Dedicated to my son, William

—Shan Liu

Dedicated to my parents and grandparents for their love and support

—Min Zhang

Dedicated to my parents for their love, encouragement, and support

—Pranav Kadam

Dedicated to my wife, Terri, and my daughter, Allison

—C.-C. Jay Kuo

Preface

Three-dimensional (3D) point clouds are gaining increasing attention for the emerging applications of 3D vision. Point clouds have widespread use in several spectrums of fields, include robotics, 3D graphics, autonomous driving, virtual reality, and so on. To keep pace with the increasing applications, the research and development of methods and algorithms to effectively store, process, and infer meaning from point cloud is on the rise. The traditional algorithms for analyzing point clouds focus on encoding the local geometric properties of points. The success of deep learning methods for processing image data led to similar networks being developed for point clouds. Present day research heavily involves the development of deep networks for various point cloud processing tasks.

The aim of this book is to give a high-level overview of point clouds and acquaint the reader with some of the most popular methods and techniques for point cloud processing. The ideal audience are those with a basic knowledge of linear algebra, machine learning, and deep learning algorithms, who wish to explore point clouds in their career or as a hobby.

This book is organized into five chapters. Chapter 1 introduces 3D point clouds and various related tasks. Chapter 2 discusses traditional point cloud analysis, including some basic operations such as filtering, nearest neighbor searching, and model fitting techniques, along with feature detectors and descriptors. Chapter 3 on deep learning discusses some of the most common machine learning-based methods. The deep learning literature is abundant, with more research being published as we write this book. We discuss some of the most representative methods that summarize the overall research direction. The emphasis is on understanding the model architecture and the novelty. The experimental details are omitted, and only key results from papers are provided. Chapter 4 on explainable machine learning methods presents our own research, which is based on a new machine learning paradigm called successive subspace learning (SSL). SSL offers several advantages over deep learning methods. Enough background review on SSL is provided prior to a thorough discussion of SSL-based methods for point cloud processing. Some applications of SSL to other vision tasks are also discussed. The final chapter (Chap. 5) includes a summary and some concluding remarks as well as possible future research directions.

Contents

Author Biographies

Shan Liu received the B.Eng. degree in Electronic Engineering from the Tsinghua University and the M.S. and Ph.D. degrees in Electrical Engineering from the University of Southern California. She is currently a Distinguished Scientist at the Tencent and General Manager of the Tencent Media Lab. She was the former Director of Media Technology Division at MediaTek, USA. She was also formerly with the MERL and Sony, etc. Dr. Liu has been actively contributing to international standards for more than a decade. She has numerous technical proposals adopted into various standards, such as H.266/VVC, H.265/HEVC, OMAF, DASH, MMT, and PCC, and served as co-editor of H.265/HEVC SCC and H.266/VVC. Meanwhile, technologies and products developed by her and her team have served hundreds of millions of users. Dr. Liu holds more than 200 granted patents and has published more than 100 technical papers. She was named "APSIPA Industrial Distinguished Leader" by the Asia-Pacific Signal and Information Processing Association in 2018, and "50 Women in Tech" by the Forbes China in 2020. She is on the Editorial Board of IEEE Transactions on Circuits and Systems for Video Technology (2018–present) and received the Best AE Award in 2019 and 2020. Her research interests include audio-visual, volumetric, immersive, and emerging media compression, intelligence, transport, and systems.

Min Zhang received her B.E. degree from the School of Science, Nanjing University of Science and Technology, Nanjing, China, and her M.S. degree from the Viterbi School of Engineering, University of Southern California (USC), Los Angeles, USA, in 2017 and 2019, respectively. She joined the Media Communications Laboratory (MCL) in the summer of 2018 and is currently a Ph.D. student in the USC, guided by Prof. C.-C. Jay Kuo. Her research interests include point cloud processing and analysis-related problems, such as point cloud classification, registration, and segmentation and detection, in the field of 3D computer vision, machine learning, and perception.

Pranav Kadam received his M.S. degree in Electrical Engineering from the University of Southern California, Los Angeles, USA, in 2020, and the Bachelor's degree in Electronics and Telecommunication Engineering from the Savitribai Phule Pune University, Pune, India, in 2018. He is currently pursuing the Ph.D. degree in Electrical Engineering from the University of Southern California. He is actively involved in the research and development of methods for point cloud analysis and processing. His research interests include 3D computer vision, machine learning, and perception.

C.-C. Jay Kuo received the Ph.D. degree in Electrical Engineering from the Massachusetts Institute of Technology, Cambridge, in 1987. He is currently the holder of William M. Hogue Professorship, a Distinguished Professor of Electrical and Computer Engineering and Computer Science, and the Director of the USC Multimedia Communications Laboratory (MCL) at the University of Southern California. Dr. Kuo is a Fellow of the American Association for the Advancement of Science (AAAS), the Institute of Electrical and Electronics Engineers (IEEE), the National Academy of Inventors (NAI), and the International Society for Optical Engineers (SPIE). He has received several awards for his research contributions, including the 2010 Electronic Imaging Scientist of the Year Award, the 2010–2011 Fulbright-Nokia Distinguished Chair in Information and Communications Technologies, the 2011 Pan Wen-Yuan Outstanding Research Award, the 2019 IEEE Computer Society Edward J. McCluskey Technical Achievement Award, the 2019 IEEE Signal Processing Society Claude Shannon-Harry Nyquist Technical Achievement Award, the 2020 IEEE TCMC Impact Award, the 72nd annual Technology and Engineering Emmy Award (2020), and the 2021 IEEE Circuits and Systems Society Charles A. Desoer Technical Achievement Award.

Chapter 1
Introduction

Abstract 3D point clouds have gained widespread attention in recent years due to their importance in 3D computer vision. In this chapter, we briefly describe the fundamentals of 3D point clouds, starting with a formal definition and the process by which they are formed. We then introduce several other popular 3D representations like 3D meshes and voxel grids, and compare these representations with point clouds. We further discuss the common tasks encountered in point cloud processing, including point cloud registration, object classification, semantic segmentation, object detection, and point cloud odometry. Next, we present some common applications of point clouds. Finally, we present some of the datasets that are frequently used in the research and development of 3D point cloud processing methods and algorithms. Overall, this introductory chapter forms the basis for the further chapters, which delve deeper into point cloud processing methods and related techniques.

1.1 Introduction

Vision is one of the five basic human senses. The human visual system scans the environment using the eyes and undertakes a series of cognitive processes in the brain to provide visual perception. Computer vision is the scientific discipline of replicating the human visual system in machines to grant them the ability of gaining understanding and reasoning from visual information like digital images and videos. Computer vision lies at the intersection of several fields, including signal and image processing, mathematics, and more recently machine learning and artificial intelligence. Several early works in computer vision focused on developing methods and algorithms for shape analysis, edge detection, pattern classification, facial recognition, and so on. The past decade has witnessed unprecedented growth in the research and development of tools and technologies for computer vision tasks. Several new applications have also emerged, such as the intersection of vision and natural language processing.

One branch of computer vision that is of particular interest to us is three-dimensional (3D) vision, and specifically 3D point cloud processing. Unlike

S. Liu et al., *3D Point Cloud Analysis*,
https://doi.org/10.1007/978-3-030-89180-0_1

traditional two-dimensional (2D) vision, which relies purely on images and projects the 3D world onto a 2D plane, 3D vision techniques provide a complementary approach by utilizing different forms of data such as depth maps, multi-view images, voxel grids, meshes, and point clouds.

1.2 3D Point Clouds

A 3D point cloud is a set of points in three-dimensional space. The point cloud represents the surface of an object in 3D. Each point consists of three coordinates that uniquely identify its location with respect to three mutually orthogonal axes. Optionally, additional information such as RGB color values and surface normal can be embedded as point attributes, depending upon the sensor used to capture the point cloud. Typically, a point cloud comprises a large number of points (tens to hundreds of thousands). Unlike 2D images, which can be represented by a regular grid, 3D point clouds are unorganized with no particular order. This unordered nature needs to be considered when dealing with point clouds and designing methods for processing them. This, combined with the large number of points, makes the realization of real-world point-cloud-based systems challenging.

1.2.1 Point Cloud Formation

The formation of 3D point clouds is very different to that of 2D images. Forming a 2D image is an optical phenomenon, in which light rays from the environment are captured using a lens, which produces an inverted image on a 2D image plane. In contrast, 3D point clouds are commonly acquired using LiDAR sensors. LiDAR stands for Light Detection and Ranging. It is closely related to its more familiar counterparts, RADAR and SONAR. Where RADAR and SONAR use radio waves and sound waves, respectively, LiDAR uses light waves. LiDAR consists of three main components: an emitter, rotator, and photodetector.

- Emitter—The emitter emits a high intensity laser beam. The wavelength of light is outside the visible spectrum to avoid interference with normal vision. Usually a wavelength in the near-infrared band is selected. The light travels in a straight line until it hits an object (obstacle). On hitting the object, the light is reflected back in the same direction (neglecting some minor effects like diffraction around the surface of the object).
- Rotator—Light traveling only in one direction is of little use. It is desirable to capture information about objects to the left, right, front, and back—in short, all directions. To achieve this, the emitter sits on a rotating device that continuously rotates. This allows the emitter to send laser beams in all possible directions. This activity happens at high speed. Coupled with the speed of the laser beam (speed

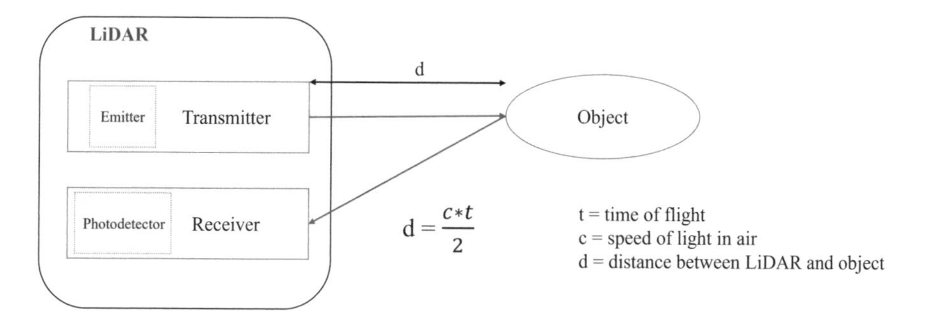

Fig. 1.1 Working principle of LiDAR

of light), huge numbers of emissions and reflections occur within a fraction of a second.
- Photodetector—The light reflected back from an object needs to be captured and analyzed to gain information about the structure of the environment. The reflected light is therefore collected by a photodetector. The detector is made from a photosensitive material, usually silicon, and is highly sensitive to the particular wavelength of light in use. The detector generates a small current when a reflected pulse hits its surface.

The operating principle of LiDAR is very simple. The emitter sends out a laser beam, which is reflected whenever it hits an object. The reflected light is sensed by the photodetector. The time instance at which the light is emitted and the time at which it is received back are noted. The difference in these two times gives the total time traveled by the laser beam. In this time, the laser beam travels twice the distance between the LiDAR sensor and the object. In this way, the distance between the LiDAR and object can be calculated using the time difference and the speed of light. The distance and direction information is recorded to produce a point in 3D space. Several thousands of these beams are emitted at the same time, which gives rise to many points. This is how point clouds are formed. The working principle of LiDAR is illustrated in Fig. 1.1.

1.2.2 Comparison with Other Visual Data Forms

Point clouds are inherently different from 2D images. While 2D images are only projections of the 3D world on a 2D plane, point clouds give 3D information about the structure of object. Point clouds are more robust under different lighting conditions and even reliable in darkness. However, one advantage of 2D images is that they have a well-defined grid structure. In contrast, point clouds are unordered in nature, which means that the methods and algorithms used to process them must be invariant under any permutation of the point cloud data and insensitive to the order in which the points appear in the stored data. The 2D grid representation in images permits the use of operations like convolution with kernels. The use of convolutional neural networks (CNNs), which are based on such convolutions, has offered tremendous benefit to 2D vision problems in recent times. However, due to the irregular structure of point clouds, such convolutions are not possible directly. In most practical cases, point clouds are very sparsely populated and contain outliers and noise.

Other forms of 3D representations include depth images (RGB-D), voxel grids, and meshes. RGB-D images combine conventional 3-channel color images (RGB) with an additional depth channel. The D value represents the distance between the image plane and the corresponding object. A popular example of a depth camera is Microsoft Kinect, which uses an infrared projector to obtain depth information. The 3D point cloud can be recovered from the RGB-D image if the camera's intrinsic parameters are known. RGB-D images can still be processed using normal convolutions, since they preserve an ordered structure like 2D images.

Voxel grids are 3D occupancy cubes that can be produced from point cloud data. Voxel grids consist of fixed-sized grids that loosely resemble pixels in 2D images. This offers a structured volumetric representation and allows 3D convolutions to be performed. However, the conversion process from point cloud to voxel grid is time consuming; hence, this representation may not be suitable to represent large-scale point clouds.

Meshes are geometric structures that approximate 3D surfaces. A single standalone point in a point cloud has little information to convey. It only makes sense when points within a neighborhood are jointly considered. These points represent discrete samples of a continuous surface. Meshes approximate this surface by fitting polygons, with the original 3D points forming the vertices.

1.3 Point Cloud Processing

We now turn our attention to some of the fundamental tasks associated with 3D point cloud processing. We begin with 3D registration followed by classification and semantic segmentation. These tasks will be discussed in depth throughout the book. Some auxiliary tasks like object detection and point cloud odometry are discussed in this section; however, the methods and algorithms pertinent to these tasks are outside the scope of this book.

1.3.1 Registration

3D registration or alignment is the process of finding a spatial transformation that optimally aligns two point clouds. Spatial transformations can be further divided into rigid and non-rigid transformations. Rigid transformations are distance-

Fig. 1.2 Point cloud indoor scene registration

preserving, whereby the distance between any two points remains the same after the transformation. 3D rotation and translation fall under this category. Non-rigid transformations, such as scaling, perspective, and affine transformations, introduce different forms of deformations. For most practical applications, the emphasis will be on rigid transformations.

The goal of registration is to find the rotation and translation that best aligns two point clouds with respect to a chosen frame of reference. The two point clouds to be registered are commonly referred to as the source and target in point cloud literature. In practice, multiple point cloud scans are registered to obtain a complete view of the environment. Each pair of point clouds typically has only a small overlapping region. The registered point cloud is then used for further processing. Registration of two point clouds from an indoor scene is shown in Fig. 1.2.

Several established methods exist for point cloud registration. We can loosely categorize these methods into correspondence-based and correspondence-free registration. The first category establishes point correspondences between the two point clouds and then uses the correspondence information to estimate the best 3D transformation. The general procedure to find correspondences is to first extract some local feature descriptors of points and then use a nearest neighbor rule. The local feature descriptors can be handcrafted or learned from data. These techniques will be discussed in detail in the subsequent chapters.

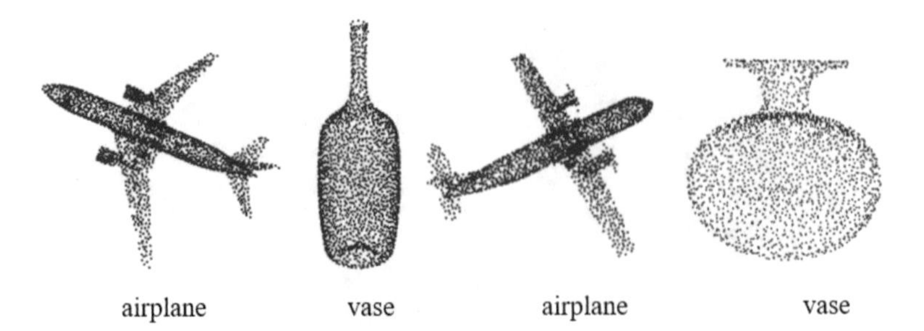

Fig. 1.3 Point cloud classification example

1.3.2 Classification

Classification is a widely used task in machine learning. From a point cloud sense, classification aims to categorize an object represented by a set of 3D points into one of the predefined classes. The general pipeline for point cloud classification is to extract global feature vectors for the point cloud and then train a machine learning classifier to recognize the object. The features can be handcrafted or learned from training data. Most recent methods train end-to-end deep networks that combine the feature extraction and classification processes. A two-class object classification example is illustrated in Fig. 1.3. The class categories are labeled below the objects.

1.3.3 Semantic Segmentation

In semantic segmentation, a class label is assigned to each point in a point cloud. The aim is to acquire a more accurate point-level understanding of the objects that constitute the point cloud and their relationships in terms of relative positions and locations. This helps to localize objects before classifying them. Figure 1.4 shows the semantic segmentation of a point cloud scan of an outdoor scene. Further, object detection for point clouds aims to surround objects of interest with 3D bounding boxes.

1.3.4 Odometry

Odometry is the process of determining the position and orientation of an object as it traverses its environment. This helps to localize objects in an incremental manner. To keep track of the position and orientation of objects, sensors like GPS, inertial measurement unit (IMU), and cameras can be used. When only camera information is used to determine the position and orientation of an object, it is called

Fig. 1.4 Semantic segmentation of an outdoor scene. Top: input point cloud, bottom: output after segmentation

visual odometry. Recently, point clouds have been increasingly used to supplement RGB images for odometry tasks. This is known as LiDAR odometry. Point cloud registration is an essential part of LiDAR odometry algorithms. Odometry is often carried out as part of bigger tasks such as simultaneous localization and mapping (SLAM), which is commonly encountered in robotics.

1.4 Applications

Some of the popular applications of point clouds include 3D graphics, animation and rendering, augmented and virtual reality, depth estimation, robotic vision, drone mapping, autonomous driving, restoration of old structures, and quality inspection in manufacturing.

Previously, automotive manufacturers primarily relied on RADAR and cameras for the development of automatic driver assistance systems (ADAS). With the advent of self-driving cars, LiDAR sensors are commonly used alongside these sensors. LiDAR point clouds enable a computer to gain a fast and reliable perception of an environment, which is critical for self-driving cars. LiDAR can accurately detect objects and measure distances to obstacles, among several related tasks. This enables fully autonomous vehicles to adapt their speed, change lanes, brake, and respond appropriately to events on the road.

Simultaneous localization and mapping (SLAM) is a fundamental task in robotics. It allows a robot to navigate an unknown environment by creating a map of its surroundings. LiDAR is a common choice of sensor for this mapping process. The process involves aligning point cloud scans captured at different positions within the environment. This facilitates further tasks such as robot path planning.

1.5 Datasets

Public datasets are critical when it comes to the research and development of algorithms. Datasets provide a common ground for evaluating algorithms and methods and enable the performance of different methods to be compared. In this section, we discuss the highlights of some commonly used datasets for different point cloud tasks. These datasets will be frequently mentioned throughout the book.

1.5.1 ModelNet40

The ModelNet40 dataset [7] is a compilation of 12,308 Computer-aided Design (CAD) models of point clouds of common objects, such as tables, chairs, sofas, airplanes, and so on. In all, the ModelNet40 dataset includes 40 object categories. The dataset is divided as 9840 models for training and 2468 models for testing. Each point cloud consists of 2048 points. All the point cloud models are pre-aligned into a canonical frame of reference. ModelNet40 and its subset ModelNet10 are widely used in point cloud object classification and shape retrieval tasks. ModelNet40 is a synthetic dataset. Some point cloud models from the ModelNet40 dataset are shown in Fig. 1.5.

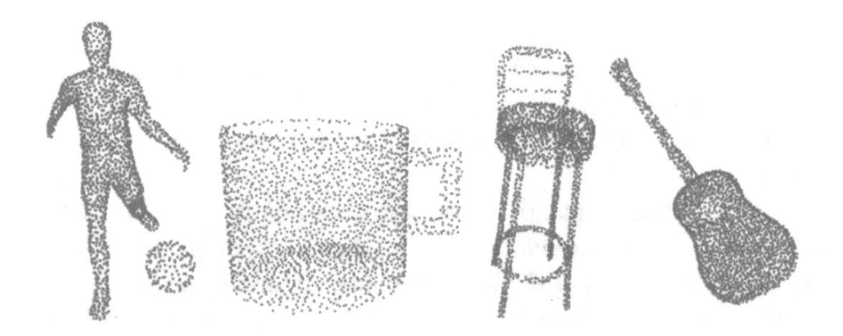

Fig. 1.5 ModelNet40 dataset. From left to right: person, cup, stool, and guitar

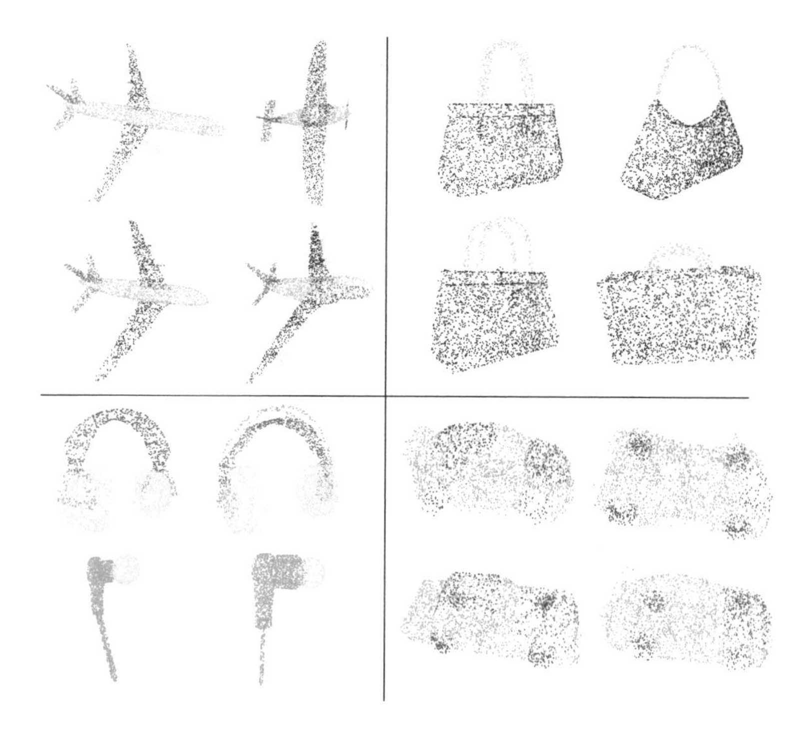

Fig. 1.6 Examples of ShapeNetPart point clouds with annotations

1.5.2 ShapeNet

The ShapeNet core dataset [2] contains 57,448 CAD models of man-made objects (airplane, bag, car, chair, etc.) in 55 categories. Each CAD model is sampled to 2048 points with three Cartesian coordinates. The ShapeNet core dataset is not fully annotated. The ShapeNetPart dataset [9] is a subset of the ShapeNet core dataset to predict a part category for each point. The ShapeNetPart dataset has 16,881 CAD models in 16 object categories, which are each sampled at 2048 points to generate point clouds. Each object category is annotated with two to six parts, and there are 50 parts in total. The dataset is divided into three sections: 12,137 shapes for training, 1870 shapes for validation, and 2874 shapes for testing. Different models of airplane, bag, earphone, and car with annotations are shown in Fig. 1.6.

1.5.3 S3DIS

The Stanford 3D Indoor Segmentation (S3DIS) dataset [1] is a subset of the Stanford 2D-3D-Semantics dataset. It is one of the benchmark datasets for point cloud semantic segmentation tasks. The S3DIS dataset contains point clouds scanned from

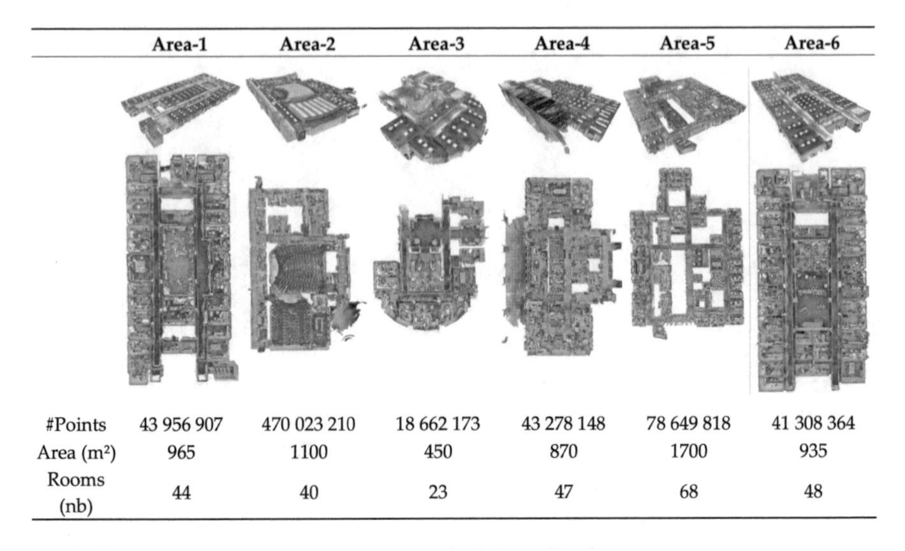

	Area-1	Area-2	Area-3	Area-4	Area-5	Area-6
#Points	43 956 907	470 023 210	18 662 173	43 278 148	78 649 818	41 308 364
Area (m²)	965	1100	450	870	1700	935
Rooms (nb)	44	40	23	47	68	48

Fig. 1.7 Details of S3DIS Dataset from [5]. Permitted by CC BY 4.0 License

6 indoor areas with 271 rooms. There are 13 categories in total, such as ceiling, floor, wall, door, etc. Each point has 9 dimensions including XYZ, RGB, and normalized XYZ. Different from ShapeNet, the dataset is labeled by object categories instead of object part categories. The dataset is usually pre-processed by block partitioning. That is, each room is split into 1×1 m blocks, where each block is randomly sampled to 4096 points for training, while all points can be used for testing depending on the memory of the computing devices. The K-fold strategy is used for training and testing (Fig. 1.7).

1.5.4 3D Match

The 3D Match dataset [10] is an ensemble of different indoor scenes from the SUN3D [8] and 7-Scenes dataset [6]. It comprises several indoor scenes such as kitchen, bedroom, office, lab, etc. This dataset is used for geometric registration of 3D indoor scenes. Each scene is constructed from 50 depth frames. The authors have used the correspondences from 3D reconstruction datasets of SUN3D and 7-Scenes to generate the ground truth labels for training. Two point clouds from this dataset are shown in Fig. 1.8.

1.5.5 KITTI

The KITTI vision benchmark suite [3] is a dataset developed for autonomous driving purposes. The KITTI dataset is a collection of RGB images, depth maps, and point

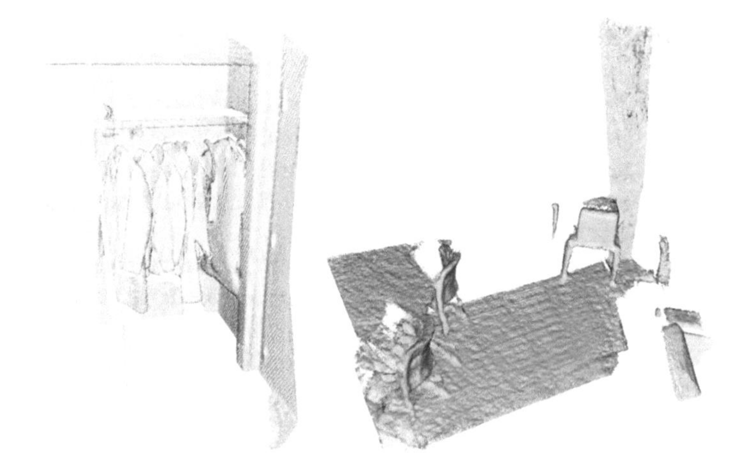

Fig. 1.8 Example of point clouds from 3DMatch dataset

Fig. 1.9 Examples of point clouds from KITTI dataset

clouds of urban street scenes in Germany, closely resembling the environment for a self-driving car. The KITTI vision challenge consists of several computer vision tasks for autonomous driving, such as stereo, odometry, object detection, tracking, semantic segmentation, and so on. One of the tasks is visual odometry/SLAM. For this task, the data consists of 22 stereo sequences, 11 with ground truth information for training and the remaining 11 for testing. The monocular/stereo images can be jointly used along with the point cloud scans for visual odometry tasks. The point clouds are captured using a Velodyne LiDAR. Two point cloud scans from the KITTI dataset are shown in Fig. 1.9.

1.6 Summary

This brings us to the end of the first chapter. Now that we have gained a basic level of insight into point clouds, we move on to more in-depth discussion on point cloud processing methods. We divide the methods into three broad categories: traditional methods, deep learning-based methods, and methods based on our own research on explainable machine learning for point clouds.

The traditional methods are based on some form of handcrafted features and the local geometric properties of points. No learning or large training data is involved. More recently, data-driven methods for point clouds have emerged. The success of deep learning—particularly CNNs—for 2D computer vision tasks has led researchers to develop similar networks for 3D point clouds. However, unlike 2D images, the unstructured and unordered nature of 3D point clouds makes it challenging to port such networks directly for point cloud processing. These deep learning-based methods require massive amounts of training data (likely labeled data in a supervised learning setting). In general, deep learning methods outperform traditional methods. The methods based on our own research on explainable machine learning for point clouds are based on the design principles of successive subspace learning (SSL) [4, 11] which offers a complementary approach to conventional deep learning-based design. More details regarding SSL and its benefits are discussed in Chap. 4.

The next chapter is on traditional methods for analyzing point clouds.

References

1. Armeni, I., Sener, O., Zamir, A.R., Jiang, H., Brilakis, I., Fischer, M., Savarese, S.: 3d semantic parsing of large-scale indoor spaces. In: Proceedings of the IEEE International Conference on Computer Vision and Pattern Recognition (2016)
2. Chang, A.X., Funkhouser, T., Guibas, L., Hanrahan, P., Huang, Q., Li, Z., Savarese, S., Savva, M., Song, S., Su, H., et al.: ShapeNet: an information-rich 3d model repository (2015). arXiv preprint arXiv:1512.03012
3. Geiger, A., Lenz, P., Stiller, C., Urtasun, R.: Vision meets robotics: the KITTI dataset. Int. J. Robot. Res. (2013)
4. Kuo, C.C.J., Zhang, M., Li, S., Duan, J., Chen, Y.: Interpretable convolutional neural networks via feedforward design. J. Vis. Commun. Image Represent. **60**, 346–359 (2019)
5. Poux, F., Billen, R.: Voxel-based 3d point cloud semantic segmentation: unsupervised geometric and relationship featuring vs deep learning methods. ISPRS Int. J. Geo-Inform. **8**(5), 213 (2019)
6. Shotton, J., Glocker, B., Zach, C., Izadi, S., Criminisi, A., Fitzgibbon, A.: Scene coordinate regression forests for camera relocalization in RGB-D images. In: Proceedings of the IEEE Conference on Computer Vision and Pattern Recognition, pp. 2930–2937 (2013)
7. Wu, Z., Song, S., Khosla, A., Yu, F., Zhang, L., Tang, X., Xiao, J.: 3d ShapeNets: a deep representation for volumetric shapes. In: Proceedings of the IEEE Conference on Computer Vision and Pattern Recognition, pp. 1912–1920 (2015)

8. Xiao, J., Owens, A., Torralba, A.: SUN3D: A database of big spaces reconstructed using SfM and object labels. In: Proceedings of the IEEE International Conference on Computer Vision, pp. 1625–1632 (2013)
9. Yi, L., Kim, V.G., Ceylan, D., Shen, I.C., Yan, M., Su, H., Lu, C., Huang, Q., Sheffer, A., Guibas, L.: A scalable active framework for region annotation in 3d shape collections. ACM Trans. Graph. **35**(6), 1–12 (2016)
10. Zeng, A., Song, S., Nießner, M., Fisher, M., Xiao, J., Funkhouser, T.: 3dmatch: learning local geometric descriptors from RGB-D reconstructions. In: CVPR (2017)
11. Zhang, M., You, H., Kadam, P., Liu, S., Kuo, C.C.J.: PointHop: An explainable machine learning method for point cloud classification. IEEE Trans. Multimedia **22**(7), 1744–1755 (2020)

Chapter 2
Traditional Point Cloud Analysis

Abstract Point cloud data is widely used in the fields of computer-aided design (CAD), augmented and virtual reality (AR/VR), robot navigation and perception, and advanced driver assistance systems (ADAS). However, point cloud data is sparse, irregular, and unordered by nature. In addition, the sensor typically produces a large number (tens to hundreds of thousands) of raw data points, which brings new challenges, as many applications require real-time processing. Hence, point cloud processing is a fundamental but challenging research topic in the field of 3D computer vision. In this chapter, we will first review some basic point cloud processing algorithms for filtering, nearest neighbor search, model fitting, feature detection, and feature description tasks. We generate some images using an open-source library, Open3D, to help illustrate the algorithms. Next, we will go over some classical pipelines for object recognition, segmentation, and registration tasks.

2.1 Filtering

3D point cloud data commonly contains noise due to sensor or environmental factors. This noise can result in inaccurate calculations regarding the local point cloud characteristics and discrepancies in the following processing steps. In addition, the large number of raw data points generated by 3D point cloud sensors means that downsampling is usually adopted to reduce the workload. In this section, we will introduce the use of filtering processes for downsampling and noise removal.

2.1.1 Downsampling

Voxel Grid Downsampling Voxel grid downsampling first builds a 3D voxel grid on top of the point cloud. Then, one point is taken from each cell to approximate all the points in that cell. There are three ways to determine which point will be taken: random selection, center point, and centroid point. The random selection approach randomly selects a point in the cell, less accurate but faster than the other methods.

S. Liu et al., *3D Point Cloud Analysis*,
https://doi.org/10.1007/978-3-030-89180-0_2

The center point approach uses the barycenter of the cell to represent the entire cell; however, this may not be a real point. The centroid approach finds the nearest point in the cell to the barycenter. The centroid approach is slightly slower than the first two approaches, but it preserves the underlying surface more accurately.

Given a point cloud $p_1, p_2, \cdots, p_n$, we first compute the minimum and maximum values for x, y, and z coordinates:

$$\begin{aligned} x_{\min} &= \min(x_1, x_2, \cdots, x_n), \\ x_{\max} &= \max(x_1, x_2, \cdots, x_n), \\ y_{\min} &= \min(y_1, y_2, \cdots, y_n), \\ &\vdots \\ z_{\max} &= \max(z_1, z_2, \cdots, z_n). \end{aligned} \tag{2.1}$$

For a voxel grid of size r, the dimensions of the voxel grid are

$$\begin{aligned} N_x &= (x_{\max} - x_{\min})/r, \\ N_y &= (y_{\max} - y_{\min})/r, \\ N_z &= (z_{\max} - z_{\min})/r, \end{aligned} \tag{2.2}$$

and the voxel index for each point is

$$\begin{aligned} i_x &= \lfloor (x - x_{\min})/r \rfloor, \\ i_y &= \lfloor (y - y_{\min})/r \rfloor, \\ i_z &= \lfloor (z - z_{\min})/r \rfloor, \\ i &= i_x + i_y * N_x + i_z * N_x * N_z. \end{aligned} \tag{2.3}$$

Subsequently, points with the same voxel index are selected based on the chosen approach (random/center/centroid point sampling). An example of voxel grid downsampling is shown in Fig. 2.1. A point cloud of cup in the 3D space is shown in the leftmost of the figure. For visualization, we set the depth of the cup to be 0

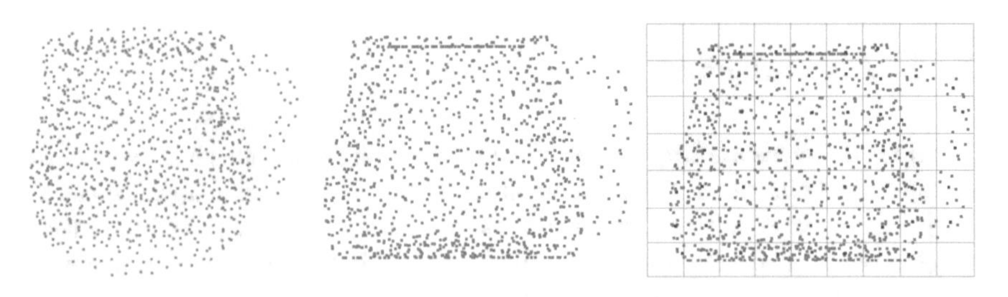

Fig. 2.1 Voxel grid downsampling by centroid point sampling approach

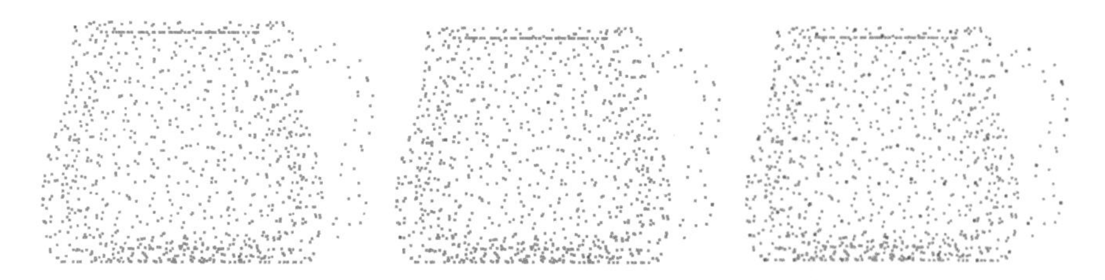

Fig. 2.2 Farthest point sampling after m iterations

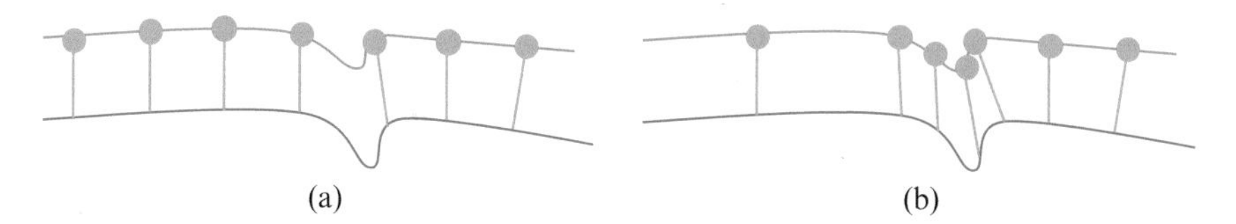

Fig. 2.3 Comparison between uniform sampling and NSS. (**a**) Uniform sampling. (**b**) NSS

so that the cup becomes 2D as in the middle of the figure. The centroid of each grid (marked in red) is taken as the sampling point.

Farthest Point Sampling (FPS) FPS [16, 22, 29] is an iterative algorithm that assigns higher possibilities to points with longer distances to the selected points until convergence. An example of FPS is shown in Fig. 2.2. A random point is first selected as the first FPS point (marked in red at the bottom of the cup) in the leftmost figure. For each of the remaining points, the distance to the nearest FPS point is calculated. The point with the largest distance is selected as the next FPS point (marked in red at the top left of the cup) in the leftmost figure. m iterations are conducted, where m is the number of points that we want to keep. The final result is shown in the rightmost of the figure.

Normal Space Sampling (NSS) NSS [33] selects points according to the surface normal. It first constructs a set of bins in the normal space, with points added into the bin according to their surface normals. Then, points are uniformly selected from all bins until the desired number of points is reached. The NSS method is used in iterative closest point (ICP) [6] for registration. A comparison between uniform sampling and NSS is shown in Fig. 2.3. With consideration of surface normal, the selection of points is more uniform.

Surface Normal Surface normal is the vector perpendicular to the tangent plane of the surface at a given point. It is widely used in plane detection and segmentation/clustering tasks, and as a feature of point clouds in deep learning. To compute the surface normal of a point p, we first find its neighborhood and fit a surface $ax + by + cz + d = 0$ over the local neighborhood. Mathematically, the surface normal will be

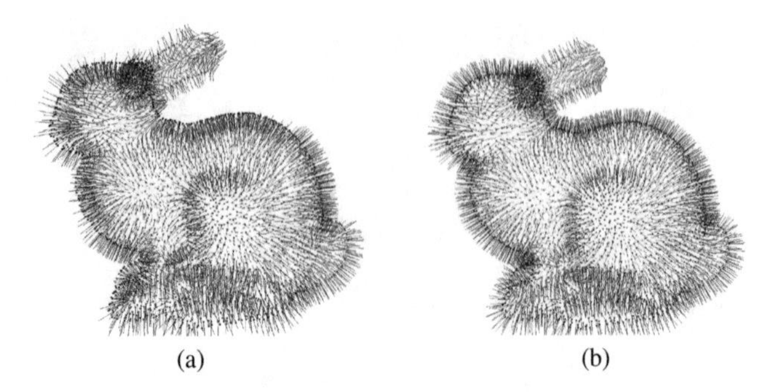

Fig. 2.4 Surface normal estimation from Open3D [46]. Copyright © 2018–2021, www.open3d.org. (**a**) Surface normal estimation. (**b**) Propagating the normal orientation

$$\mathbf{n} = [n_x, n_y, n_z]^T = \frac{[a, b, c]^T}{\|[a, b, c]^T\|}. \tag{2.4}$$

More generally, we can apply principle component analysis (PCA) [41] on the neighborhood to obtain eigenvalues λ_1, λ_2, and λ_3, where $\lambda_1 >= \lambda_2 >= \lambda_3$, and the corresponding eigenvectors v_1, v_2, and v_3. The least significant component, v_3, is the surface normal n. That is,

$$\mathbf{n} = v_3. \tag{2.5}$$

The proof is omitted here. An example of surface normal estimation by PCA is shown in Fig. 2.4. The right-hand figure panel uses a minimum spanning tree to propagate the normal orientation.

In addition to the surface normal, the eigenvalues and eigenvectors of the local structure tensor are commonly used in constructing geometric features. For example, the surface variation (i.e., curvature c) is the ratio between eigenvalues:

$$c = \frac{\lambda_3}{\lambda_1 + \lambda_2 + \lambda_3}. \tag{2.6}$$

Other geometric features are shown in Fig. 2.5, including covariance-based, moment-based, and height-based features.

2.1.2 Noise Removal

Radius Outlier Removal Radius outlier removal filters points in a cloud based on the number of neighbors in a certain radius. For each point, we first find its neighbors

covariance	Sum	$\lambda_1 + \lambda_2 + \lambda_3$
	Omnivariance	$(\lambda_1 \cdot \lambda_2 \cdot \lambda_3)^{\frac{1}{3}}$
	Eigenentropy	$-\sum_{i=1}^{3} \lambda_i \cdot \ln(\lambda_i)$
	Anisotropy	$(\lambda_1 - \lambda_3)/\lambda_1$
	Planarity	$(\lambda_2 - \lambda_3)/\lambda_1$
	Linearity	$(\lambda_1 - \lambda_2)/\lambda_1$
	Surface Variation	$\lambda_3/(\lambda_1 + \lambda_2 + \lambda_3)$
	Sphericity	λ_3/λ_1
	Verticality	$1 - \lvert\langle[0\,0\,1], \mathbf{e}_3\rangle\rvert$
moment	1st order, 1st axis	$\sum_{i\in\mathcal{P}}\langle \mathbf{p}_i - \mathbf{p}, \mathbf{e}_1\rangle$
	1st order, 2nd axis	$\sum_{i\in\mathcal{P}}\langle \mathbf{p}_i - \mathbf{p}, \mathbf{e}_2\rangle$
	2nd order, 1st axis	$\sum_{i\in\mathcal{P}}\langle \mathbf{p}_i - \mathbf{p}, \mathbf{e}_1\rangle^2$
	2nd order, 2nd axis	$\sum_{i\in\mathcal{P}}\langle \mathbf{p}_i - \mathbf{p}, \mathbf{e}_2\rangle^2$
height	Vertical range	$z_{\max} - z_{\min}$
	Height below	$z - z_{\min}$
	Height above	$z_{\max} - z$

Fig. 2.5 Geometric features based on eigenvalues of the local structure tensor [20]. Permitted by CC BY 3.0 License

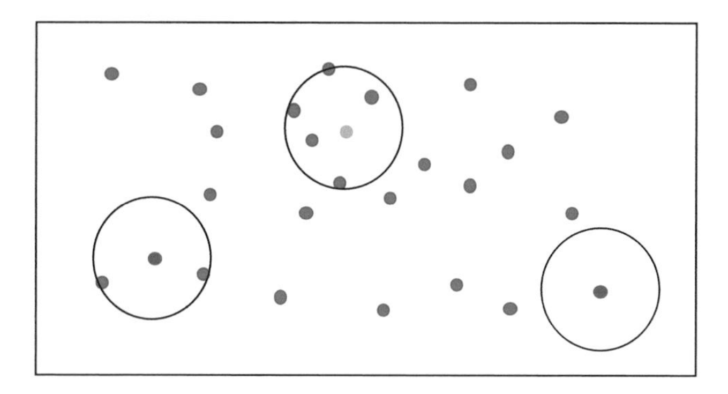

Fig. 2.6 Radius outlier removal: $k_{min} = 4, r = 1$

within a certain radius r. If the number of neighbors k is less than a given number k_{min}, it will be considered as an outlier and removed. An example of radius outlier removal is shown in Fig. 2.6. The green point is considered an inliner, while the red points are outliers.

Statistical Outlier Removal Statistical outlier removal removes points that are further away from their neighbors. For each point, we first find a neighborhood and calculate its distance to the neighbors d_{ij} considering the point index $i = 1, \cdots, n$ and the neighbor index $j = 1, \cdots, k$. Then, the distances are modeled by Gaussian distribution $d \sim N(\mu, \sigma)$:

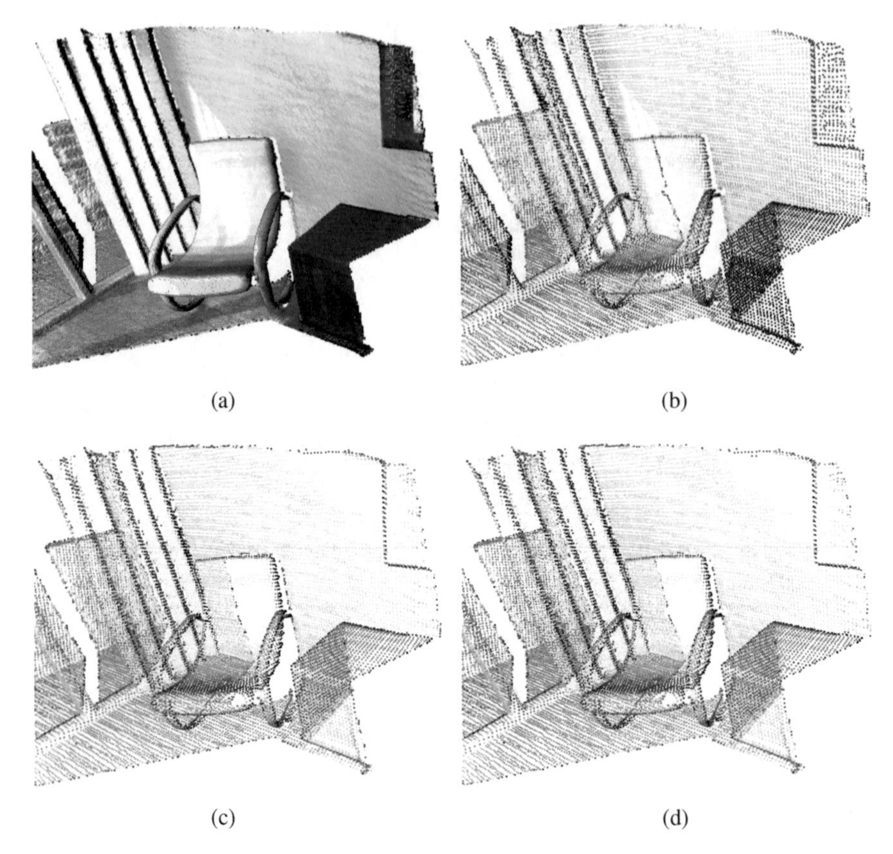

Fig. 2.7 Summary of point cloud filtering methods from Open3D [46]. Copyright © 2018–2021, www.open3d.org. (**a**) Original point cloud. (**b**) Voxel grid downsampling: voxel size 0.02. (**c**) Radius outlier removal: $k_{min} = 16, r = 0.05$. (**d**) Statistical outlier removal: $k = 20, c = 2$

$$\mu = \frac{1}{nk}\sum_{i=1}^{n}\sum_{j=1}^{k} d_{ij}, \quad \sigma = \sqrt{\frac{1}{nk}\sum_{i=1}^{n}\sum_{j=1}^{k}(d_{ij} - \mu)^2}. \tag{2.7}$$

If the mean distance to the neighboring points is outside a defined interval, the point will be removed. For example, the point will be removed if

$$\frac{1}{k}\sum_{j=1}^{k} d_{ij} > \mu + c\sigma \quad \text{or} \quad \frac{1}{k}\sum_{j=1}^{k} d_{ij} < \mu - c\sigma, \; c \in \mathbb{R}^{+}. \tag{2.8}$$

A point cloud collected from scanning device is shown in Fig. 2.7a; it is then downsampled by voxel grid downsampling with grid size 0.02 in Fig. 2.7b.

Then, radius outlier removal and statistical outlier removal are conducted on the downsampled point cloud in Fig. 2.7c and d, respectively. The outliers are marked in red, and the inliers are marked in gray.

2.2 Nearest Neighbor Search

As we mentioned above, clouds have an irregular structure. Whereas the local neighborhood of pixels in 2D images can be easily identified by creating a grid around the pixel, point clouds do not have natural grid-based representation, and it is nontrivial to build grids. Instead, nearest neighbor (NN) search [23] serves as a basic element for building local neighborhoods for points in a point cloud. NN search is used in radius noise removal, statistical noise removal, and surface normal estimation tasks, as described in Sect. 2.1, to calculate local characteristics for each point with its local neighborhood. Moreover, it will frequently appear in the following contents of this book.

The nearest neighbor problem has two main solutions: K-NN [1, 19] and fixed radius-NN [4, 5] (Fig. 2.8). Given N points in a space S, for a query point $p \in S$, K-NN finds the K closest points to p in space S, while radius-NN finds all points q that satisfy the criteria $\|q - p\| < r$. The radius-NN approach is slightly faster than the K-NN approach, because the time complexity of finding distances that are smaller than r is $O(N)$ for each point while finding top K smallest distances for each point takes from $O(N\log K)$ to $O(N\log N)$ according to the sorting algorithm. But K-NN is more accurate than radius-NN. The algorithm itself is straightforward. However, the problem becomes complex when the huge data throughput of real-time applications is considered. For example, Velodyne HDL-64E scans 2.2 million points per second (110,000 points at 20 Hz); the brute-force radius-NN algorithm's time complexity is $O(N^2)$, which makes $110{,}000 \times 110{,}000 = 12 \times 10^9$ comparisons at 20 Hz. It is unavoidable to consider more efficient ways of realizing such an algorithm for point cloud processing. Among the various solutions, space partitioning NN is commonly used. This method first splits the space into different areas, and only some areas are searched instead of all the data. In this section, three data structures are used to illustrate this: binary

Fig. 2.8 Comparison of K-NN and Radius-NN. (**a**) K-NN. (**b**) Radius-NN

search tree (BST) [11], k-dimensional (k-d) tree [3], and octree [28]. BST is for one-dimensional (1D) data, k-d tree works for data of any dimension, and octree is optimized for 3D data.

2.2.1 Binary Search Tree

Binary Search Tree (BST) is a node-based tree data structure. Each node stores a key. Each key is greater than the keys in the left subtree of the node but lower than the keys in the right subtree of the node. The left and right subtrees are both BSTs. The basic operations of BST include tree construction, insertion, searching, and deletion.

BST Construction We first construct a BST using all the points in the point cloud. This can be completed by recursively inserting elements. The worst-case time complexity is $O(h)$ where h is the height of the BST. If the tree is extremely imbalanced (i.e., a chain), then h is the number of points. If the tree is balanced, then $h = \log_2 N$.

Example of 1-NN Search by BST An example is shown in Fig. 2.9 to explain 1-NN search by BST. Assuming the query point is key 11, we want to find its nearest neighbor with the constructed BST by traversing the tree. The traversal process is described briefly below:

- Starting at the root node (key 8), the worst distance is $|11 - 8| = 3$. Because all points in the left subtree are 3 or more away from 11, we only need to search the right subtree;
- Go to the node with key 10, and update the worst distance as $|11 - 10| = 1$, because $1 < 3$. Then, search the right subtree of node 10;
- Go to the node with key 14, the distance is $|11 - 14| = 3$. Because $3 > 1$, the worst distance remains 1. The right subtree of node 14 will therefore only be farther away, while the left subtree may reduce the distance. Thus, the left subtree of node 14 is then searched;
- Go to the node with key 13, the distance is $|11 - 13| = 2$. Because $2 > 1$, the worst distance is still 1.
- When there are no more subtrees, the search ends.

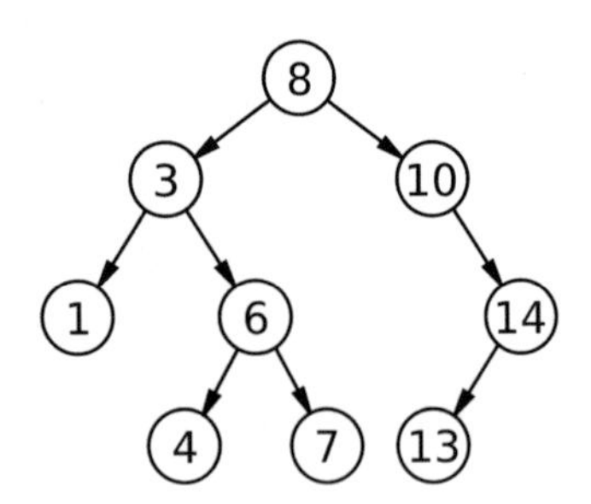

Fig. 2.9 Binary search tree (BST) from [12]. Public domain image

K-NN Search by BST The K-NN search by BST is similar to the example of 1-NN search, except for the worst distance. In K-NN search, a container is built to store the K nodes, and the worst distance is the largest distance of nodes in the store to the query node which is dynamic. The areas outside the worst circle can be skipped, while the areas inside the worst circle are searched and the worst distance is updated.

Radius-NN Search by BST The radius-NN search by BST is similar to the K-NN search by BST in that it skips the areas that are outside the worst circle. The only difference is that the worst distance is fixed rather than dynamic, which is a fixed radius we set.

BST Search Time Complexity BST-based NN search shortens the time complexity for brute-force NN search from $O(N)$ to $O(\log N)$ for one query point if the tree is balanced, where N is the number of points.

2.2.2 *k-Dimensional Tree*

A k-dimensional (k-d) tree is a binary tree where every leaf node is a k-dimensional point. It is an extension of BST into higher dimensions, invented by Jon Louis Bentley in 1975. Examples of 2D and 3D k-d trees are shown in Fig. 2.10.

k-d Tree Construction k-d tree construction, starting from the root node, takes place as follows:

- If the node has only one point or the number of points is less than the leaf size, stop splitting and store the node as a leaf node;

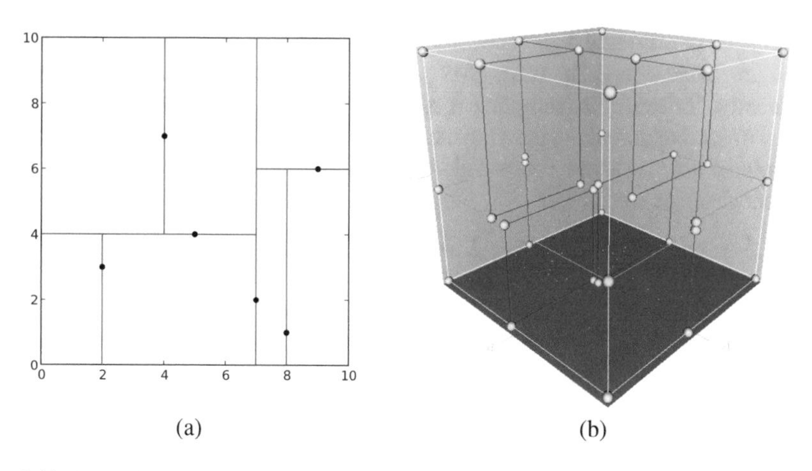

Fig. 2.10 Examples of 2D and 3D k-d trees from [8]. The splitting position is one of the points. Permitted by GNU General Public License. (**a**) 2D k-d tree. (**b**) 3D k-d tree

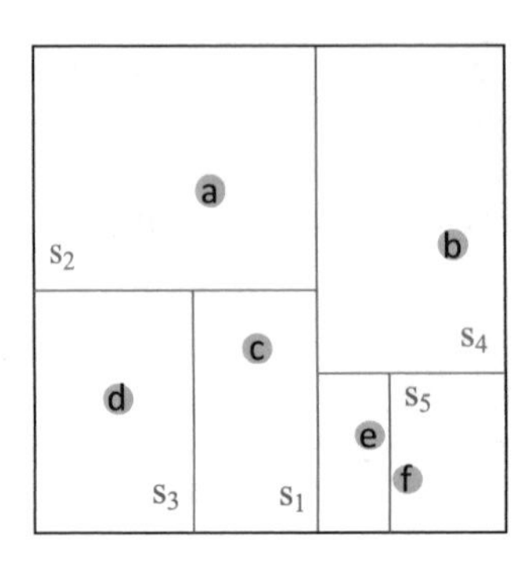

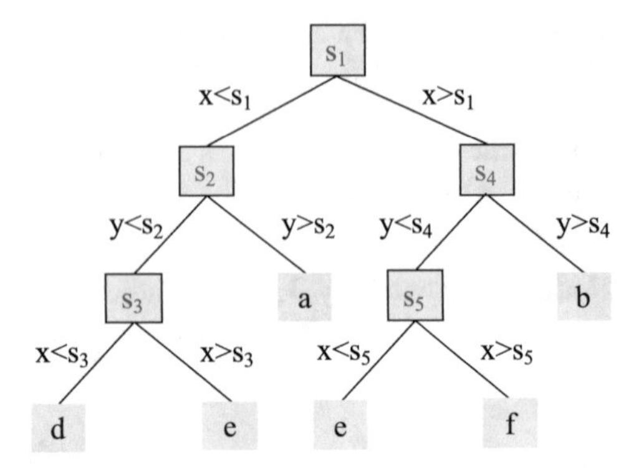

Fig. 2.11 Example of 2D k-d tree construction

- Otherwise, divide the points of the node into two parts by a hyperplane perpendicular to the selected splitting axis. Points to the left of the hyperplane go to the left subtree of that node, and points to the right of the hyperplane go to the right subtree;
- Repeat the first two steps until the stopping criteria are satisfied.

The splitting position can be or not be one of the points. The splitting axis can either be round-robin: $x \rightarrow y \rightarrow z \rightarrow x \rightarrow y \rightarrow \cdots$ or adaptive (the axis with widest spread).

An example of 2D k-d tree construction is shown in Fig. 2.11. To construct the 2D k-d tree, we first sort the values to obtain the median position s_1 along the x-axis and split all the points to two subtrees. This is repeated for the left and right subtrees. For example, the median position along y-axis is computed for all points in the left subtree. Iteration is continued until the stopping criteria are met. The time complexity is $O(\log N(N \log N))$ due to the sorting at each level of the tree. With a $O(N)$ median finding approach, the k-d tree can be constructed in $O(N \log N)$. There are some simple methods that work well in practice but have lower time complexity, for example, sampling a subset of the points in each node for sorting instead of sorting all points, or using the mean instead of the median. However, these methods cannot guarantee a balanced k-d tree where each leaf node is approximately the same distance to the root.

***K*-NN Search by *k*-d Tree** The K-NN search by k-d tree is conducted from the root node to the leaf node that covers the query point. The query point is compared to all points in the leaf node. Then, the tree is traversed accordingly. Up to K points are stored, and the worst distance is updated each time. The criteria for when a partition intersects with the worst distance are as follows:

- The splitting axis is inside the partition.
- The distance between the splitting axis and the partition is less than the worst distance.

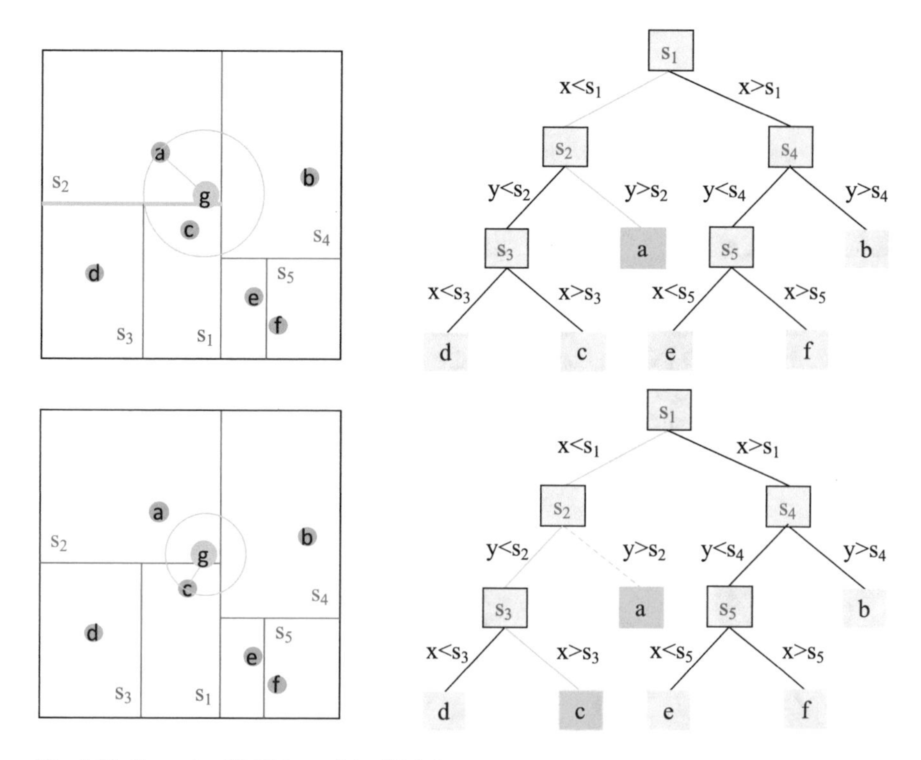

Fig. 2.12 Example of 2-NN search by 2D k-d tree

A 2-NN search by k-d tree example is shown in Fig. 2.12. In this case, the initial worst distance d is infinite. The search follows the following sequence:

- The left-side of s_1 and up-side of s_2 is the partition that covers the query node g, so point a is stored. The worst distance is still infinite.
- Since the distance $|s_2 - g_y| < d$, go to the left subtree of s_2 and right subtree of s_3, so point c is stored. The worst distance now becomes $d = \|g - a\|$ since $\|g - a\| > \|g - c\|$.
- Since $|s_1 - g_x| < d$ and $|s_4 - g_y| > d$, go to the right subtree of s_4, but because $\|g - b\| > d$, b is not stored.

Radius-NN Search by k-d Tree Radius-NN search by k-d tree is similar to that for K-NN search. The only difference is that the worst distance is fixed rather than dynamic. That is, the worst distance is the fixed radius that we set.

k-d Tree Search Time Complexity The time complexity of 1-NN search in a balanced k-d tree is $O(\log N)$ for one query point. The time complexity for K-NN or radius-NN search can vary from $O(\log N)$ to $O(N)$, depending on the distribution of points and K or r. However, it is nontrivial to determine whether the NN search by k-d tree is complete, so we must return to the root every time.

2.2.3 *Octree*

Octrees are a tree data structure wherein each node has eight children nodes as shown in Fig. 2.13. It is most commonly used to partition 3D space into eight octants recursively, which is analog of quadtree [17] for partitioning 2D space into four quadrants. Octant is an element/node of the octree, and it is a cube. Octrees are different from k-d trees in that k-d trees split along a dimension, while octrees split around a point; k-d trees are binary, while octrees are not.

Octree Construction There are two kinds of octree: point region (PR) octrees and matrix based (MX) octrees. The nodes of PR octrees store the center of the region as a pseudo 3D point that further defines one of the corners of each of the eight children. The nodes of MX octrees are implicitly the center of the space they represent. A 2D example of a PR octree construction is shown in Fig. 2.14. We first

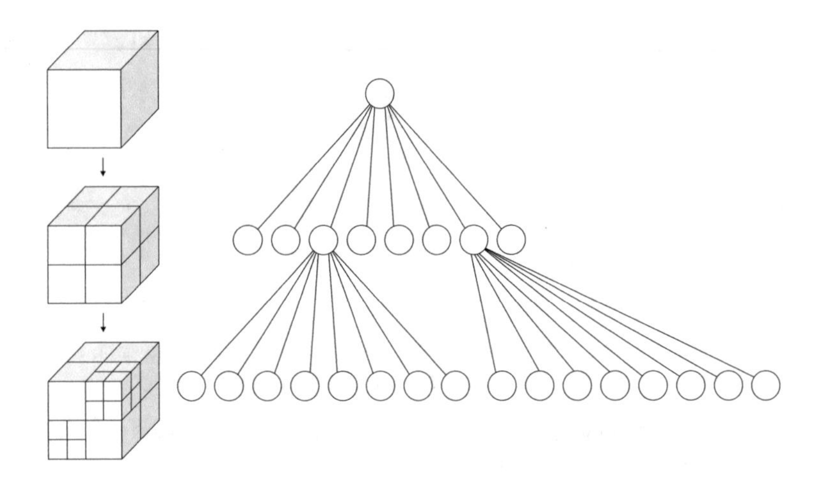

Fig. 2.13 Octree from [40]. Permitted by CC BY-SA 3.0 Unported License

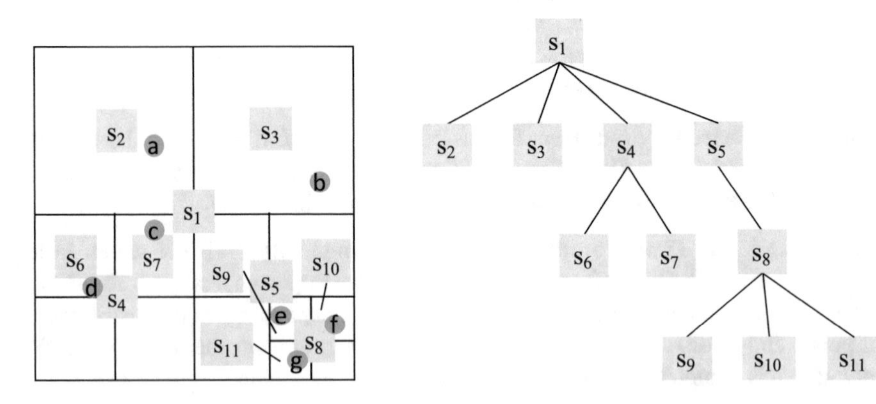

Fig. 2.14 2D Octree construction

determine the extent of the first octant s_1 and then determine whether to further split the octant by checking the leaf size, minimum extent and maximum iteration, and so on. The minimum extent is to avoid infinite splitting when there are repeated points. s_1 is then split into s_2, s_3, s_4, and s_5. s_2 and s_3 have only one point, so they are leaf nodes. s_4 can be further divided into s_6 and s_7, and the empty octant is discarded. s_5 repeat the above procedures.

***K*-NN Search by Octree** The K-NN search by octree uses a depth-first search (DFS) [11] method. It searches from the root octant to the leaf octant that contains the query point and then compares with all the points in the leaf and stores up to K points. From this, the worst distance is updated, which forms a query ball. In every iteration, the most relevant child is determined and searched first, followed by other children. If the octant overlaps with the query ball, all points in the leaf are compared to update the K points, worst distance, and query ball. If the query ball is inside the octant, search ends.

An example of 2-NN search by octree is shown in Fig. 2.15. In this case, the initial worst distance d is infinite. The search follows the following sequence:

- Starting from s_1 to s_2, which contains the query point h, point a is stored and the distance between h and a forms a query ball.

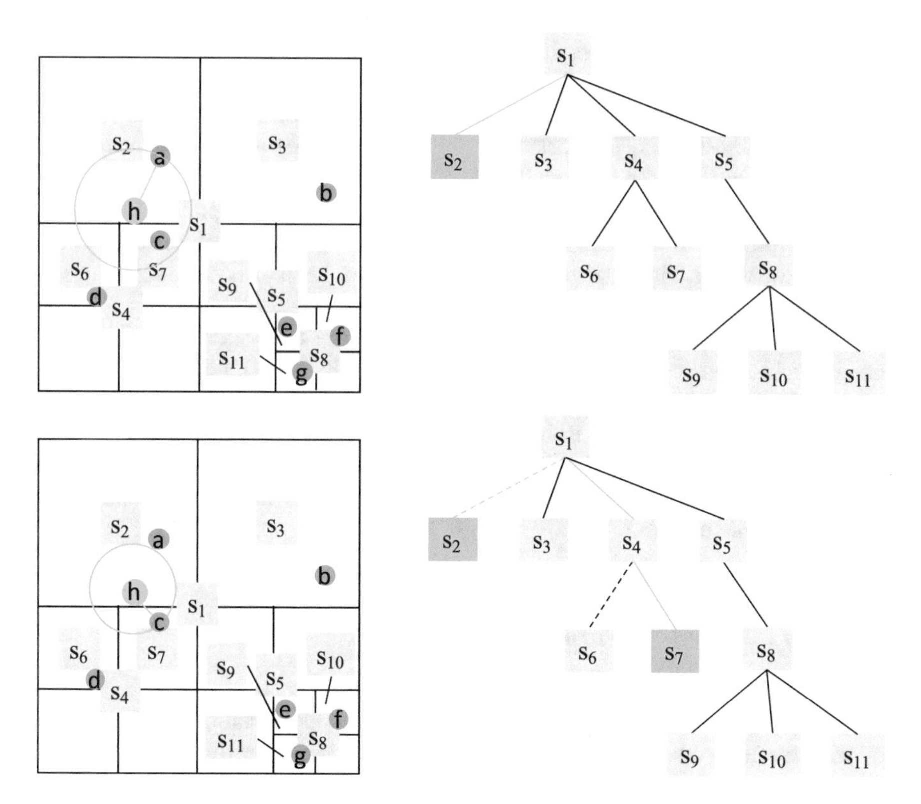

Fig. 2.15 2-NN search by 2D octree

- Since the query ball intersects with s_4, we check the children of s_4. Point c is stored and $d = \|h - a\|$.
- Since octants s_3 and s_5 do not intersect with the query ball, the search ends.

Radius-NN Search by Octree A simple method to conduct radius-NN search by octree fixes the worst distance as the radius, while the rest procedures are the same as the K-NN search by octree. A better approach is to check if the query ball contains the octant. If the query ball contains the octant, the query can just be compared with all points in the octant and there is no need to consider the children of that octant.

Octree Search Time Complexity The time complexity of 1-NN search by octree is $O(\log N)$ for one query point. The search time complexity of K-NN or radius-NN can vary from $O(\log N)$ to $O(N)$, depending on the distribution of points and K or r. Octrees are more efficient than k-d trees, because the search can be stopped without returning to the root. By using depth-first search, the nodes to be traversed and the required surfaces to be viewed can be identified.

2.3 Model Fitting

Model fitting is a common technique for processing point cloud data. Large-scale point cloud data usually contains a large portion of data points which can be described by explicit mathematical equations, such as lines and planes. For example, in the KITTI 3D object detection dataset, the LiDAR data captures not only the objects of interest (cars, pedestrians, cyclist, etc.) but also other elements of the environment, such as the ground. The ground is less valuable than the objects of interest; however, in most scenarios, there are far more ground points than points related to objects of interest. Removing the ground points therefore reduces the computational cost. Since the ground can be considered as a plane with some noises, it can be solved by model fitting.

The three standard approaches to model fitting are least squares [25], Hough transform [14], and random sample consensus (RANSAC) [18]. Least squares work well if the inliners are known. However, even a small number of outliers will considerably affect the result. Robust least squares, Hough transform, and RANSAC can handle outliers better. Hough Transform and RANSAC are extremely robust when the number of outliers is close to or more than that of inliers or when the data include multiple models.

2.3.1 *Least Squares Fitting*

Least squares fitting is a mathematical approach for approximating the solution of overdetermined systems where the equation number is greater than the number of unknowns. It minimizes the sum of the squares of the residuals in every equation. Given a set of points $p_1, p_2, \cdots, p_n$, where $p_i \in \mathbb{R}^2$, to find a line

$$ax + by + c = 0 \tag{2.9}$$

that fits the point set best, the sum of the squares of the residuals

$$E = \sum_{i=1}^{n} (ax_i + by_i + c)^2 \tag{2.10}$$

should be minimized. Therefore, this model fitting problem is usually reformulated as standard linear least squares (LLSQ) optimization for $A\mathbf{x} = 0$:

$$\hat{\mathbf{x}} = \min_{\mathbf{x}} \|A\mathbf{x}\|^2, \ A \in \mathbb{R}^{n \times m}, \ \mathbf{x} \in \mathbb{R}^m. \tag{2.11}$$

Here, we have $m = 3$

$$A = \begin{bmatrix} x_1 & y_1 & 1 \\ \vdots & \vdots & \vdots \\ x_n & y_n & 1 \end{bmatrix}, \ \mathbf{x} = [a, b, c]^T,$$

$$\hat{\mathbf{x}} = [\hat{a}, \hat{b}, \hat{c}]^T = \min_{\mathbf{x}} \|A\mathbf{x}\|^2, \ s.t. \ \|x\|_2 = 1.$$

Given A is full column rank, i.e., $n \geq 3$, the solution is easily found by finding the eigenvector of the smallest eigenvalues of A.

Besides, many model fitting problems can be formulated as LLSQ optimization problems for $A\mathbf{x} = \mathbf{b}$:

$$\hat{\mathbf{x}} = \min_{\mathbf{x}} \|A\mathbf{x} - \mathbf{b}\|^2, \ A \in \mathbb{R}^{n \times m}, \ \mathbf{x} \in \mathbb{R}^m, \ \mathbf{b} \in \mathbb{R}^n. \tag{2.12}$$

Similarly, given full column rank A,

$$\hat{\mathbf{x}} = (A^T A)^{-1} A^T \mathbf{b}. \tag{2.13}$$

The main limitation of LLSQ, which restricts its application in point cloud processing, is its sensitivity to outliers. Ordinary least squares estimation is optimal only if no outliers exist in the data. As is shown in Fig. 2.16, a single outlier results in poor line fitting using the ordinary least squares method. Typically, the weight assigned to each observation is expected to be $1/n$; however, the outliers exert a far greater weight than they deserve. Therefore, outliers pull the fit too far in their direction. In most cases, it is difficult to identify the problem, because the residuals of the outliers are much smaller than they otherwise would be. Robust least squares require less restrictive assumptions. It reduces the weights of outliers to increase

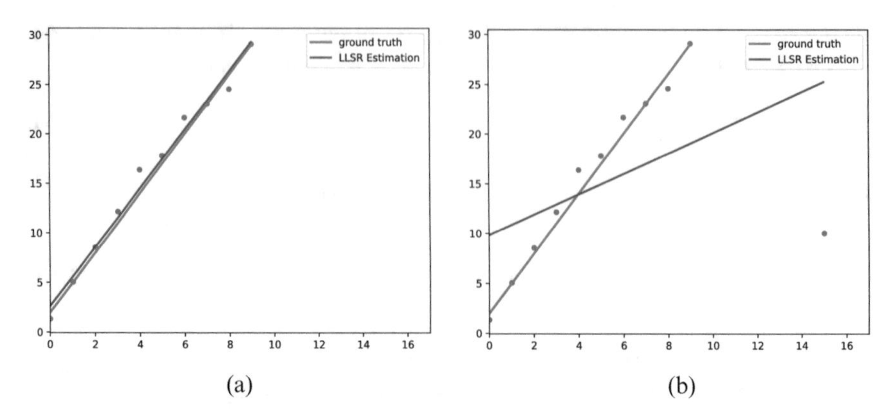

Fig. 2.16 Ordinary least squares are sensitive to outliers. (**a**) Only inliers. (**b**) One outlier added

their residuals. Therefore, the influence of outliers is decreased so that most data can be fit well. It is also easier to identify outliers than in ordinary least squares.

Ordinary least squares minimize the sum of the squares of the residuals, i.e., the L_2-norm of the residuals. This is also known as L_2-norm regression. An alternative is robust least squares. For example, we can use the L_1 loss function instead, which minimizes the sum of the absolute of the residuals, i.e., the L_1-norm of the residuals. Compared with L_2-norm regression, L_1-norm regression assigns a smaller weight to the outliers, thereby reducing their influence. Typical loss functions include:

$$\begin{aligned}
&L_1. && \rho = |s|, \\
&L_2. && \rho = s^2, \\
&\text{Cauchy.} && \rho = \log(1 + |s|), \\
&\text{Huber.} && \rho = \begin{cases} s^2, & |s| < \delta \\ 2\delta(|s| - \frac{1}{2}\delta), & \text{otherwise} \end{cases} \\
&etc.
\end{aligned} \tag{2.14}$$

Cauchy and Huber are robust loss functions, which reduce the effect of outliers; however, the problem becomes nonlinear. An example of robust least squares is shown in Fig. 2.17, which is much better than ordinary least squares in Fig. 2.16b.

A general formulation of LSQ is

$$\hat{\mathbf{x}} = \min_{\mathbf{x}} \| f(\mathbf{x}) \|^2. \tag{2.15}$$

Here, function f can be linear or nonlinear function. In LLSQ, f is a linear combination of parameters, while in nonlinear least squares (NLLSQ), the parameters appear as functions. The model is linear if the derivatives of f to the parameters

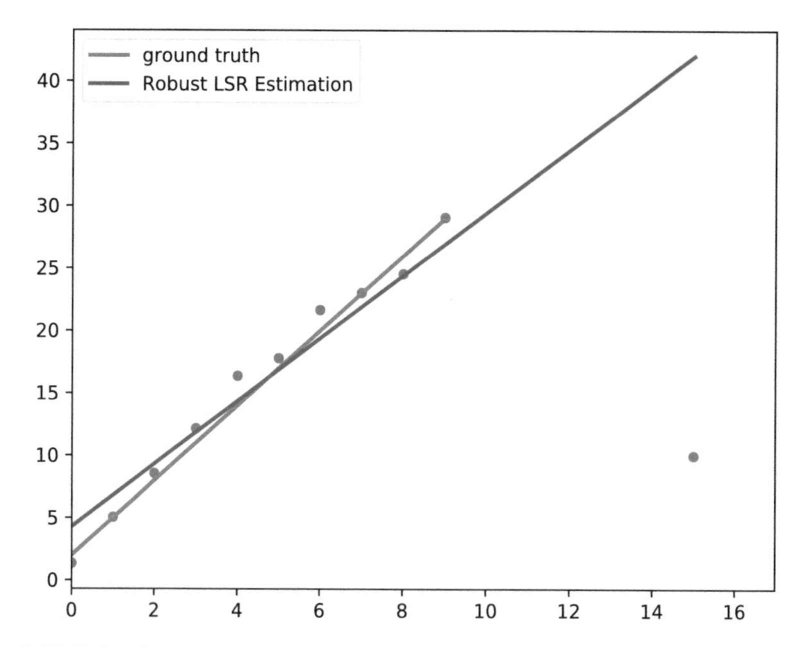

Fig. 2.17 Robust least squares using Huber function

are constant or depend only on the values of the independent variable; otherwise, it is nonlinear. LLSQ is globally convex, so it can be solved analytically to achieve a unique solution. In contrast, the analytical solution to NLLSQ problems requires the use of partial derivatives to the Jacobians, which can be complicated. It is common to use optimization methods like gradient descent, Gauss–Newton, and Levenberg–Marquardt to solve NLLSQ problems iteratively. Furthermore, there may be multiple minima in NLLSQ problems; non-convergence is a common phenomenon in NLLSQ.

2.3.2 Hough Transform

The Hough transform is a feature extraction technique that can be used to isolate features of a particular shape within an image. The classical Hough transform is most commonly used to detect lines, circles, or ellipses, which can be specified by some parametric form. A generalized Hough transform extends to the detection of features that cannot be expressed analytically. The Hough transform first maps the data from the image space to the parameter space and then uses a voting method to obtain the solution. The analytical equation for a straight line is

$$y = ax + b, \tag{2.16}$$

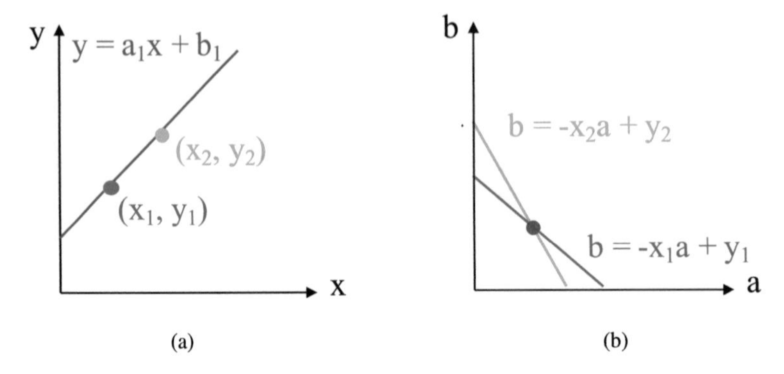

Fig. 2.18 Hough transform. (**a**) Image space. (**b**) Hough space

which corresponds to a single point (a, b) in the parameter space. In addition, a point (x, y) in the Euclidean space corresponds to the line

$$b = -xa + y \tag{2.17}$$

in the parameter space. The mapping between image space and Hough space is shown in Fig. 2.18.

In the model parameterization of line $y = ax + b$, a is infinity for vertical lines, which is not uniform in the parameter space. A better model with parameters (θ, r) is usually adopted instead:

$$x\cos\theta + y\sin\theta = r. \tag{2.18}$$

Consider fitting a line to a set of discrete points in an image; the Hough transform first generates a line in the parameter space for each data point. Ideally, these lines will intersect to give a single solution, e.g., the red point in Fig. 2.18b. However, they will not intersect if there is outlier. Therefore, the Hough transform uses voting to solve the problem. The parameter space is discretized into bins, which are also known as accumulator space, and the intersections of lines are viewed as votes. Then, the local maximum, i.e., the bin with the most votes, is selected.

As an extension, the Hough transform can be used to detect circles

$$(x - a)^2 + (y - b)^2 = r^2 \tag{2.19}$$

with parameters (a, b, r). Each point (x, y) is mapped to a parameter bin in the (a, b, r) space. To find the bin, we first fix $r = r_i$, and uniformly sample a set of $\{\theta_1, \cdots, \theta_k\}$. Each θ_j generates a set of $\{a, b\}$, where

$$\begin{aligned} a &= x - r_i\cos\theta_j, \\ b &= x - r_i\sin\theta_j. \end{aligned} \tag{2.20}$$

Moreover, Hough transform can be easily extended to 3D objects detection such as planes and cylinders in point clouds. A plane is represented by

$$z = ax + by + c, \tag{2.21}$$

where a 3D Hough space can be constructed from parameters (a, b, c). However, extending the Hough transform to plane detection suffers from the problem of infinite a and b when the planes become vertical, because big values amplify the noise in the data. For horizontal plane detection, it works quite well. To detect cylindrical objects in point clouds, whereby the orientation of the cylinder is first identified, followed by its position and radius.

When implementing Hough transform, a trade-off between speed and precision is achieved through the selection of resolution and the application of smoothing at the parameter space before searching for the highest vote to reduce the influence of outliers. The advantage of the Hough transform is its robustness to noise and to missing points of the shape. It can be extended to numerous models and usually works well for models with less than 3 unknown parameters. However, it does not scale well with complicated models.

2.3.3 Random Sample Consensus

RANSAC estimates the parameters of a mathematical model iteratively from a set of observations. It is non-deterministic which means the parameter estimation is reasonable within a certain probability that increases with the number of iterations.

RANSAC completes line fitting by repeating the following steps:

- Randomly select a sample that is a minimal subset of the points required to solve the model:

$$p_0 = (x_0, y_0), \ p_1 = (x_1, y_1);$$

- Solve the line model with

$$y_0 = ax_0 + b,$$
$$y_1 = ax_1 + b;$$

- Compute the error function for each point $p_i = (x_i, y_i)$

$$d_i = \frac{n^T(p_i - p_0)}{\|n\|^2},$$

  where $n = [a, b]^T$.
- Count the points that are consistent with the model, i.e., inliers where $d_i < \tau$.

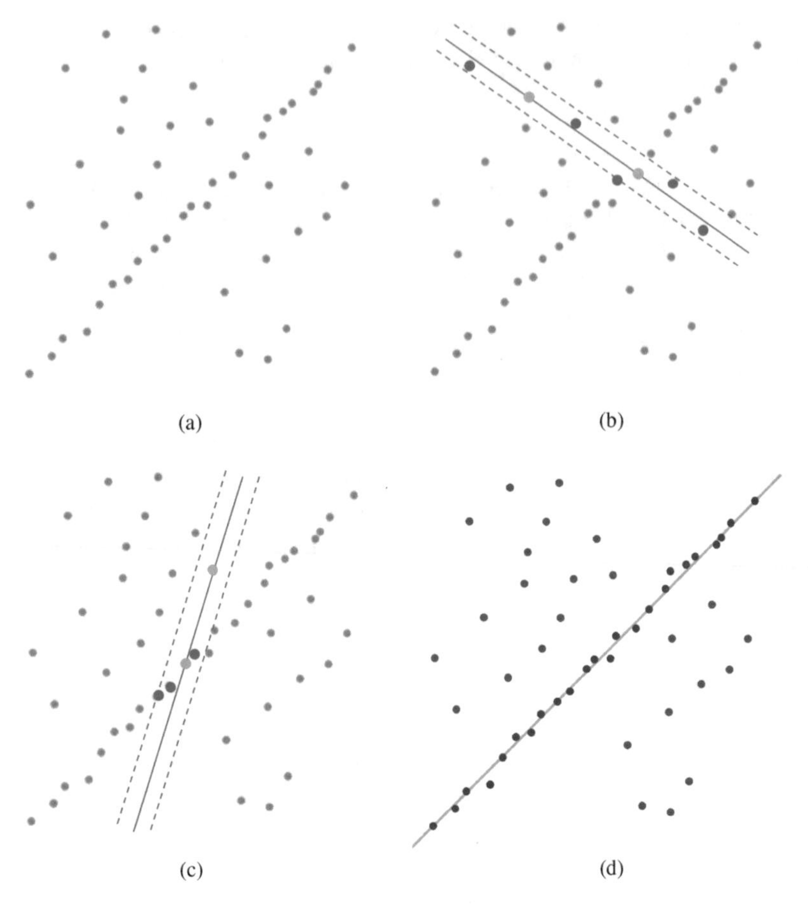

Fig. 2.19 Line fitting by RANSAC. Reproduced from [30, 31]. Permitted by CC BY-SA 3.0 License. (**a**) Data points. (**b**) First round line fitting. (**c**) Second round line fitting. (**d**) Final result

Repeat above steps for N iterations, the model with the most inlier points will be chosen as final model. Examples of several iterations are shown in Fig. 2.19. The blue points are inliers, the green points in Fig. 2.19b and c are randomly selected samples in each iteration, and the red points in Fig. 2.19d are outliers. Notably, the outliers have no influence on the result; hence, RANSAC can also be used as an outlier detection method.

In RANSAC implementation, two parameters need to be considered: the distance threshold τ and the number of iterations N. The distance threshold τ is usually chosen empirically or by Chi-square (χ^2) distribution, which is the sum of the squares of k independent standard normal distributions. Assume the error between data points and the model has a Gaussian distribution $d \sim \mathcal{N}(0, \sigma^2)$ and the point is an inlier with 95% probability, then

$$\tau = \begin{cases} \sqrt{3.84\sigma^2}, & \chi_1^2 \text{ for 2D/3D line fitting or 3D plane fitting} \\ \sqrt{5.99\sigma^2}, & \chi_2^2 \text{ for 2D point distance} \\ \sqrt{7.81\sigma^2}, & \chi_3^2 \text{ for 3D point distance} \end{cases} \tag{2.22}$$

N is chosen to let at least one random sample be free from outliers with probability p, e.g., $p = 0.99$. Assume the probability that a point is an outlier is e so that the probability of choosing s inliers is $(1-e)^s$, where $s = 2$ in line fitting. Hence, the probability that we will not obtain any good sample from N iterations is

$$1 - p = (1 - (1 - e)^s)^N. \tag{2.23}$$

That is,

$$N = \frac{\log(1 - p}{\log(1 - (1 - e)^s)}. \tag{2.24}$$

As is shown in Table. 2.1, as the proportion of outliers and minimum number of points needed to solve the model increase, the number of iterations needed also increases. In practice, several approaches can be applied to reduce the number of iterations. For example, the iteration can be stopped when the inlier ratio reaches the expected value. In addition, least squares can be used to refine the model after selecting the final model and inlier points.

RANSAC has several advantages. In particular, it is simple to use and works well with complicated models, even with small inlier ratios of around 10%. The disadvantage is that the distance ratio needs to be determined manually.

Table 2.1 Number of iterations w.r.t proportion of outliers and minimal number of required points for solving the model

	Proportion of outliers e						
Minimal required # of points s	5%	10%	20%	25%	30%	40%	50%
2	2	3	5	6	7	11	17
3	3	4	7	9	11	19	35
4	3	5	9	13	17	34	72
5	4	6	12	17	26	57	146
6	4	7	16	24	37	97	293
7	4	8	20	33	54	163	588
8	5	9	26	44	78	272	1177

2.4 Point Cloud Features

Features are elements that provide information about the content of an image or a certain region of an image in computer vision and image processing. Depending on the specific problem to solve, features may be edges, corners/interest points, and blobs/region of interest, which have a specific structure. They can be found by using general neighborhood operation or feature detection operations.

Feature detection is a low-level image processing operation that is usually performed as the first operation on an image. It computes and makes decisions at every pixel to see whether there is a feature at that pixel. The detected features are usually in the form of isolated points, continuous curves, or connected regions. Owing to the wide usage of feature detection operations in many computer vision algorithms, various feature detectors have been developed such as the Harris corner detector [21], SUSAN (Small Univalue Segment Assimilating Nucleus) corner detector [38], and Scale-Invariant Feature Transform (SIFT) [26].

After identifying a point of interest by feature detection, a feature vector is usually extracted to describe the point. This is known as feature description. Depending on the applications, a further step of feature matching may be necessary. For example, when stitching images together to produce a panorama [7], it is necessary to determine the correspondence between the descriptors of the given images. The process is shown in Fig. 2.20. Other applications of feature matching include simultaneous localization and mapping (SLAM) [2, 15].

Point cloud features operations borrow ideas from image features operations. A common application is point cloud registration, which is to find a transformation, i.e., rotation and translation, to align two point clouds. A classical method is iterative closest point (ICP) [6], which requires an initial guess. Another method is to detect, describe, and match features as in the image domain, for which no initialization is required. An example for point cloud registration is shown in Fig. 2.21.

2.4.1 Feature Detectors

There are two types of point cloud feature detector. The first borrows from image feature detectors such as Harris corners 3D/6D [37], which are extensions of image based Harris corners, while other uses intrinsic shape signatures (ISS) [44], which is a native method for point cloud processing using 3D geometry.

2.4.1.1 Harris 3D/6D

For images, corners are the points at which edges meet, and edges are a sudden change in image brightness. Corners are more distinctive than flat regions or edges. First, corners have large variations in neighborhood. While flat regions have no

Fig. 2.20 Feature detection, description, and matching for panorama. Reproduced with permission [7]. Copyright © 2006, Springer Science Business Media, LLC. (**a**) SIFT matches 1. (**b**) SIFT matches 2. (**c**) RANSAC inliers 1. (**d**) RANSAC inliers 2. (**e**) Images aligned according to a homography

changes in any direction, and edges have no change along the edge direction, corners have significant changes in all directions. Second, corners are invariant to translation, rotation, and illuminations. Although images contain a relatively small number of corners, the corners contain significant features.

Fig. 2.21 Point cloud feature detection, description, and matching for registration. Reproduced with permission [13]. Copyright © 2018, IEEE

Harris corner detectors operate based on the intensity change of a patch. Given an image I, if a patch $x, y \in \Omega$ is shifted by $[u, v]$, then the intensity change is

$$E(u, v) = \sum_{x,y\in\Omega} w(x, y)[I(x + u, y + v) - I(x, y)]^2, \tag{2.25}$$

where $w(x, y)$ is the window function, which can be binary or Gaussian. When u, v is minimized, the Taylor expansion for $I(x + u, y + v)$ is

$$\begin{aligned} I(x + u, y + v) = {} & I(x, y) + uI_x(x, y) + vI_y(x, y) \\ & + \frac{1}{2!}[u^2 I_{xx}(x, y) + uvI_{xy}(x, y) + v^2 I_{yy}(x, y)] + \cdots . \end{aligned} \tag{2.26}$$

Assume $w(x, y) = 1, \forall x, y \in \Omega$, the first order approximation of $E(u, v)$ is

$$E(u, v) \approx \begin{bmatrix} u & v \end{bmatrix} M \begin{bmatrix} u \\ v \end{bmatrix}, \quad M = \sum_{x,y\in\Omega} \begin{bmatrix} I_x^2 & I_x I_y \\ I_x I_y & I_y^2 \end{bmatrix}, \tag{2.27}$$

where M is the covariance matrix of the image gradient. The intensity gradient for pixel i at location $[x_i, y_i]$ can be denoted as $I_i = [I_{x_i}, I_{y_i}]^T$, then $M = \sum_i I_i I_i^T$. To understand the physical meaning behind I_i and M, consider three cases:

1. Linear edge: $I_i = [I_{x_i}, 0]^T$.
2. Flat region: $I_i = [0, 0]^T$.
3. Corner: $I_i = [I_{x_i}, I_{y_i}]^T$.

The eigenvalues of M are λ_1 and λ_2. λ_1 and λ_2 are both small for a flat patch and both large for a corner. For edges, $\lambda_1 >> \lambda_2$ for a vertical edge and $\lambda_2 >> \lambda_1$ for a horizontal edge. Then, a response function R is calculated for each pixel

$$R = \lambda_1\lambda_2 - k(\lambda_1 + \lambda_2)^2 = \det(M) - k(tr(M))^2, \tag{2.28}$$

where $k \in [0.04, 0.06]$ is set empirically. After normalization, pixels with $R > \tau$ are corners and τ is a threshold. Thereafter, non-maximum suppression (NMS) can be used to find the local maxima as corners within the window, which is a 3×3 filter.

Harris 3D Harris 3D is an extension of Harris corner detector for processing point clouds with/without intensity. If the point cloud has intensity, the intensity change of a small move $[u, v, w]$ for a local region Ω over a point (x, y, z) is

$$E(u, v, w) = \sum_{x,y,z\in\Omega} [I(x+u, y+v, z+w) - I(x, y, z)]^2. \tag{2.29}$$

Then, the first order approximation is

$$E(u, v, w) \approx \begin{bmatrix} u & v & w \end{bmatrix} M \begin{bmatrix} u \\ v \\ w \end{bmatrix}, \quad M = \sum_{x,y,z\in\Omega} \begin{bmatrix} I_x^2 & I_x I_y & I_x I_z \\ I_x I_y & I_y^2 & I_y I_z \\ I_x I_z & I_y I_z & I_z^2 \end{bmatrix}, \tag{2.30}$$

where M is the covariance matrix of intensity over the surface. The intensity gradient for point i can be denoted as a vector $\mathbf{e} = [e_x, e_y, e_z]^T$, the direction of which is the greatest intensity increase in a direction. Ideally,

$$(x_j - x_i)e_x + (y_j - y_i)e_y + (z_j - z_i)e_z = I(x_j, y_j, z_j) - I(x_i, y_i, z_i), \tag{2.31}$$

where $x_j, y_j, z_j \in \Omega$. In matrix form,

$$A\mathbf{e} = \mathbf{b}, \ A = \begin{bmatrix} \Delta x_1 & \Delta y_1 & \Delta z_1 \\ \vdots & \vdots & \vdots \\ \Delta x_j & \Delta y_j & \Delta z_j \\ \vdots & \vdots & \vdots \end{bmatrix}, \mathbf{b} = \begin{bmatrix} \Delta I_1 \\ \vdots \\ \Delta I_j \\ \vdots \end{bmatrix}. \tag{2.32}$$

Hence, the problem can be formulated as

$$\hat{\mathbf{e}} = \min_{\mathbf{e}} \|A\mathbf{e} - \mathbf{b}\|^2, \ A \in \mathbb{R}^{n\times 3}, \ \mathbf{e} \in \mathbb{R}^3, \ \mathbf{b} \in \mathbb{R}^n. \tag{2.33}$$

Given full column rank A, the solution is

$$\hat{\mathbf{e}} = (A^T A)^{-1} A^T \mathbf{b}, \tag{2.34}$$

which gives $[I_x, I_y, I_z]^T$ in M. As we know, point cloud data from LiDAR scans contain points that correspond to the objects of interest and the environment, so we can optionally project the intensity gradient $\mathbf{e}$ onto the local surface to reduce the effect of noise:

$$\mathbf{e}' = \mathbf{e} - \mathbf{n}(\mathbf{e}^T \mathbf{n}). \tag{2.35}$$

The eigenvalues for M, $\lambda_1, \lambda_2, \lambda_3$, are sorted in descending order. Thus, the corner response R is λ_3. Since there might be an intensity corner on the surface, λ_2 is also valid. The response selection follows the Kanade–Tomasi corner detector [36] method for more stable tracking.

If the point cloud does not have intensity information, we can assume a local surface Ω around point p is $f(x, y, z) = 0$. A cost function can be constructed similarly:

$$E(u, v, w) = \sum_{x,y,z\in\Omega} [f(x+u, y+v, z+w) - f(x, y, z)]^2. \tag{2.36}$$

Then, the local surface becomes a plane with the first order approximation:

$$f(x, y, z) \approx ax + by + cz + d = 0, \tag{2.37}$$

where $[a, b, c]^T$ is the surface normal $\mathbf{n} = [n_x, n_y, n_z]^T$. The distance to the plane is

$$\begin{aligned} f(x+u, y+v, z+w) &\approx a(x+u) + b(y+v) + c(z+w) + d \\ &= [x+u, y+v, z+w]\mathbf{n}. \end{aligned} \tag{2.38}$$

Then, the first order approximation for $E(u, v, w)$ is

$$E(u, v, w) \approx \begin{bmatrix} u & v & w \end{bmatrix} M \begin{bmatrix} u \\ v \\ w \end{bmatrix}, \ M = \sum_{x,y,z\in\Omega} \begin{bmatrix} n_x^2 & n_x n_y & n_x n_z \\ n_x n_y & n_y^2 & n_y n_z \\ n_x n_z & n_y n_z & n_z^2 \end{bmatrix}, \tag{2.39}$$

where M is the covariance matrix of surface normal over the surface. The Tomasi response R is λ_3.

Harris 6D Harris 6D uses a combination of intensity and surface normal. M is the covariance matrix of $[I_x, I_y, I_z, n_x, n_y, n_z]$,

$$M = \sum_{x,y,z \in \Omega} \begin{bmatrix} I_x^2 & I_x I_y & I_x I_z & I_x n_x & I_x n_y & I_x n_z \\ I_x I_y & I_y^2 & I_y I_z & I_y n_x & I_y n_y & I_y n_z \\ I_x I_z & I_y I_z & I_z^2 & I_z n_x & I_z n_y & I_z n_z \\ n_x I_x & n_x I_y & n_x I_z & n_x^2 & n_x n_y & n_x n_z \\ n_y I_x & n_y I_y & n_y I_z & n_x n_y & n_y^2 & n_y n_z \\ n_z I_x & n_z I_y & n_z I_z & n_x n_z & n_y n_z & n_z^2 \end{bmatrix}. \tag{2.40}$$

Here, the Tomasi response R is λ_4. When $R = \lambda_3$, Harris 6D is a superset of Harris 3D without intensity which is too loose. When $R = \lambda_5$, it requires the corner to be both a geometric corner and an intensity corner, which is too strict.

2.4.1.2 Intrinsic Shape Signatures

ISS is originally a shape descriptor for 3D object recognition. Given a point $p_i \in \mathbb{R}^3$, we first find its nearest neighbors in a radius r and assign a weight w_i that is inversely related to the number of points in its local neighborhood:

$$w_i = \frac{1}{\|p_j : |p_j - p_i| < r\|}. \tag{2.41}$$

This means the points in sparsely sampled regions will contribute more than those in densely sampled regions. Then, we compute the weighted covariance matrix for each point

$$COV(p_i) = \frac{\sum_{|p_j - p_i| < r} w_j (p_j - p_i)(p_j - p_i)^T}{\sum_{|p_j - p_i| < r} w_j}. \tag{2.42}$$

Subsequently, we decompose the covariance matrix to obtain its eigenvalues $(\lambda_i^1, \lambda_i^2, \lambda_i^3)$ in the order of decreasing magnitude and their corresponding eigenvectors. Points with large variations in their local neighborhood are recognized as keypoints because they tend to contain more unique information of the shape than points in the flat regions. Keypoints can be extracted by comparing the eigenvalues, because their smallest eigenvalues, λ_i^3, should be larger than those of the other points. A flat surface has eigenvalues $\lambda_1 = \lambda_2 > \lambda_3$ and a line has eigenvalues $\lambda_1 > \lambda_2 = \lambda_3$. Hence, p_i is a keypoint if

$$\frac{\lambda_i^2}{\lambda_i^1} < \gamma_{21} \quad \text{and} \quad \frac{\lambda_i^3}{\lambda_i^2} < \gamma_{32} \tag{2.43}$$

to ensure $\lambda_1 > \lambda_2 > \lambda_3$. Then we perform NMS with λ_i^3 so that not too many points from the same region are selected.

2.4.2 *Feature Descriptors*

Once a keypoint is detected, feature descriptor can be used to extract a vector around the keypoint to describe it. There are two main types of feature descriptors: histogram-based and signature-based. The histogram-based methods encode local geometric variations into a histogram, including point feature histogram (PFH) [34] and fast point feature histogram (FPFH) [34]. The signature-based methods compute point distributions based on a local reference frame (LRF). Signature of histogram of orientations (SHOT) [39] is a representative signature-based method.

2.4.2.1 Point Feature Histogram

The aim of PFH is to capture the surface variation in a neighborhood by creating a LRF based on the surface normal to achieve 6D-pose independent. Given a point p, we first find its nearest neighbors within a radius r. A LRF is then defined for each pair of points (p_1, p_2) in the sphere. For point p_1,

$$\begin{aligned} u &= n_1, \\ v &= u \times \frac{p_2 - p_1}{\|p_2 - p_1\|}, \\ w &= u \times v, \end{aligned} \tag{2.44}$$

where n_1 is the surface normal at p_1. Then, for each pair of points, compute a quadruplet $[\alpha, \phi, \theta, d]$:

$$\begin{aligned} d &= \|p_2 - p_1\|, \\ \alpha &= u \cdot n_2, \\ \phi &= u \cdot \frac{p_2 - p_1}{\|p_2 - p_1\|}, \\ \theta &= \arctan(w \cdot n_2, u \cdot n_2) \end{aligned} \tag{2.45}$$

where n_2 is the surface normal at p_2. Since we do not want the descriptors to be dependent on the viewpoint, d is usually ignored. Therefore, there are k^2 such quadruplets/triplets for k points in the neighborhood. Next, we treat the triplet as a 3D data point and enter it into the histogram. The histogram is similar to a 3D voxel grid, whereby each dimension has B bins. The PFH feature vector is a B^3 array after flatting the voxel grid and conducting normalization.

2.4.2.2 Fast Point Feature Histogram

FPH is a simple and effective feature description method, but it has a time complexity of $O(k^2)$ for each keypoint with a local neighborhood of k points. FPFH improves the time complexity to $O(k)$. Simplified point feature histogram (SPFH) is used to only compute the triplets between query points and their neighbors within r, with each dimension entered into a histogram of B bins. Once the SPFH is computed for query point p, the SPFH of the neighboring points is computed. FPFH is the weighted sum of neighboring SPFH,

$$FPFH(p) = SPFH(p) + \frac{1}{k}\sum_{i=1}^{k} w_i \cdot SPFH(p_i),\ w_i = \frac{1}{\|p - p_i\|}. \tag{2.46}$$

As is shown in Fig. 2.22, while PFH uses fully connected neighbors within r, the neighbors in FPFH are only partially connected, with a coverage range of up to $2r$. Some edges in FPFH are counted twice. Instead of using a voxel grid, three histograms are concatenated directly. Thus, the FPFH feature descriptor has a size of $3B$.

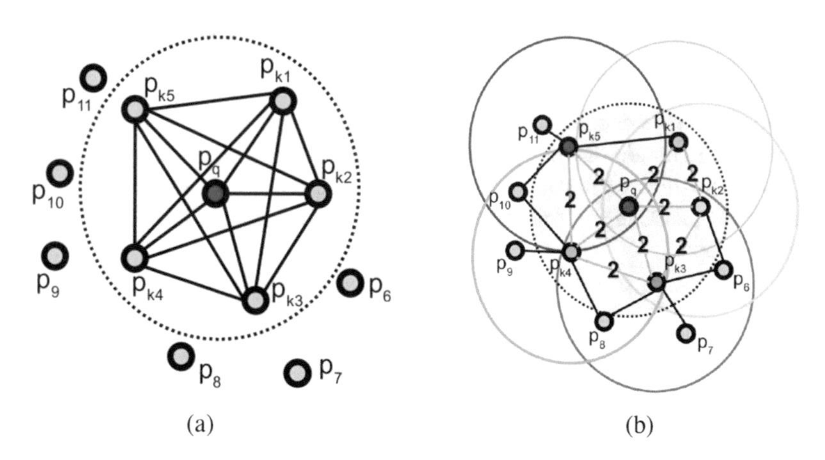

Fig. 2.22 Comparison between PFH and FPFH. Reproduced with permission [34]. Copyright © 2009, IEEE. (**a**) PFH: fully connected neighbors. (**b**) FPFH: partially connected neighbors

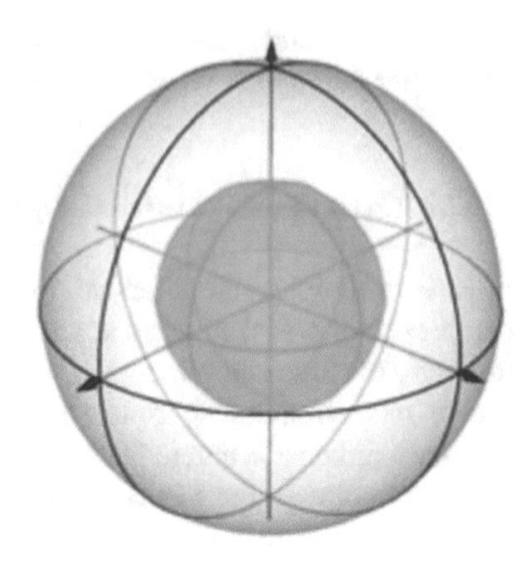

Fig. 2.23 Signature structure for SHOT: Four azimuth divisions for visualization. Reproduced with permission [39]. Copyright © 2010, Springer-Verlag Berlin Heidelberg

2.4.2.3 Signature of Histograms of Orientations

While histogram-based methods like PFH and FPFH capture surface variations in a neighborhood, the neighborhood position is not directly recorded. To encode information about the neighborhood position, SHOT (Fig. 2.23) first builds a canonical pose of the local neighborhood, i.e., the local reference frame (LRF), to ensure the local coordinate is 6D-pose independent. The weighted covariance matrix M over the local neighborhood within r is given by:

$$M = \frac{1}{\sum_{i:d_i<r}(r-d_i)} \sum_{i:d_i<r} (r-d_i)(p_i-p)(p_i-p)^T,\ d_i = \|p_i - p\|. \tag{2.47}$$

Subsequently, M is decomposed to find the eigenvectors in the decreasing order of eigenvalue magnitude. Owing to the sign ambiguity in PCA, each principal vector can have a positive or negative direction, denoted as $\mathbf{x}^+, \mathbf{y}^+, \mathbf{z}^+$ and $\mathbf{x}^-, \mathbf{y}^-, \mathbf{z}^-$. Note that $\mathbf{x}$, $\mathbf{y}$, $\mathbf{z}$ are eigenvectors here, rather than coordinates. The direction of $\mathbf{x}$ is determined by:

$$\begin{aligned} \mathbf{x} &= \begin{cases} \mathbf{x}^+, & |S_{\mathbf{x}}^+ \geq |S_{\mathbf{x}}^-| \\ \mathbf{x}^-, & \text{otherwise} \end{cases} \\ S_{\mathbf{x}}^+ &\doteq i : d_i \leq r \wedge (p_i - p) \cdot \mathbf{x}^+ \geq 0, \\ S_{\mathbf{x}}^- &\doteq i : d_i \leq r \wedge (p_i - p) \cdot \mathbf{x}^- > 0, \end{aligned} \tag{2.48}$$

where $S_{\mathbf{x}}^+$ and $S_{\mathbf{x}}^-$ are the number of points in the half space of $\mathbf{x}^+$ and $\mathbf{x}^-$, respectively. Similarly, the direction of $\mathbf{z}$ is determined. Then, $\mathbf{y} = \mathbf{z} \times \mathbf{x}$.

With the local reference frame, the space is then divided into 32 volumes by 8 azimuth divisions, 2 elevation divisions, and 2 radial divisions in the LRF. For each volume, we build a histogram of B bins for $\cos\theta_i$:

$$\cos\theta_i = n_p \cdot n_{p_i}, \tag{2.49}$$

where n_p is the surface normal of the keypoint and n_{p_i} is the surface normal of a point in the volume. The descriptor size is $32B$. Compared with PFH and FPFH, SHOT only connects a keypoint with its neighbors, yet has the same time complexity as FPFH ($O(nk)$).

The boundary effect in SHOT should be taken into consideration. That is, points at the edge of each volume should contribute to the neighboring volume as well. For a point $p_i = (\rho_i, \alpha_i, \beta_i, \cos\theta_i)$, where ρ_i is the distance to keypoint, α_i is the azimuth angle, and β_i is the elevation angle. This can contribute up to 8 volumes and 2 bins:

$$\left(\left\lfloor\frac{\rho_i}{r_\rho}\right\rfloor \text{ and } \left\lceil\frac{\rho_i}{r_\rho}\right\rceil, \left\lfloor\frac{\alpha_i}{r_\alpha}\right\rfloor \text{ and } \left\lceil\frac{\alpha_i}{r_\alpha}\right\rceil, \left\lfloor\frac{\beta_i}{r_\beta}\right\rfloor \text{ and } \left\lceil\frac{\beta_i}{r_\beta}\right\rceil, \left\lfloor\frac{\cos\theta_i}{r_\theta}\right\rfloor \text{ and } \left\lceil\frac{\cos\theta_i}{r_\theta}\right\rceil\right), \tag{2.50}$$

where $r_\rho, r_\alpha, r_\beta, r_\theta$ are division resolutions.

If the point is in volume $(\lceil\frac{\rho_i}{r_\rho}\rceil, \lfloor\frac{\alpha_i}{r_\alpha}\rfloor, \lfloor\frac{\cos\beta_i}{r_\beta}\rfloor)$ and histogram bin $\lfloor\frac{\cos\theta_i}{r_\theta}\rfloor$. The contribution of the point is $w = w_\rho w_\alpha w_\beta w_\theta$, where

$$\begin{aligned}
w_\rho &= 1 - \frac{\lceil\frac{\rho_i}{r_\rho}\rceil r_\rho - \rho_i}{r_\rho},\\
w_\alpha &= 1 - \frac{\alpha_i - \lfloor\frac{\alpha_i}{r_\alpha}\rfloor r_\alpha}{r_\alpha},\\
w_\beta &= 1 - \frac{\beta_i - \lfloor\frac{\cos\beta_i}{r_\beta}\rfloor r_\beta}{r_\beta},\\
w_\theta &= 1 - \frac{\cos\theta_i - \lfloor\frac{\cos\theta_i}{r_\theta}\rfloor r_\theta}{r_\theta}.
\end{aligned} \tag{2.51}$$

2.5 Classification and Segmentation

Segmentation is a fundamental task in 3D point cloud processing. Given a point cloud, the goal of segmentation is to cluster points with similar patterns into one category. The segmentation process further helps with the location, recognition, and classification of objects. Point cloud classification is often called semantic segmentation or point labeling in traditional point cloud perception. We can assign each segment with a class label once a point cloud has been segmented. The traditional segmentation can be categorized into five groups: edge-based, region-based, attributes-based, model-based, and graph-based methods [32].

1. Edge-based methods detect edges where the points have rapid intensity change. Those edges are usually the boundaries for different regions in the point clouds. Hence, the regions of the point cloud are segmented.
2. Region-based methods search the neighborhood first. Points in the neighborhood that have similar patterns are combined to form isolated regions, followed by finding dissimilarity between different regions.
3. Attributes-based methods first compute the attributes of point cloud data and then cluster point clouds based on the attributes.
4. Model-based methods are pure geometric. The points which have the same mathematical representation/geometric shape, e.g., spheres, cones, planes, and cylinders, are grouped as one segment.
5. Graph-based methods consider point clouds as graphs. A simple model is each vertex corresponds to a point and the edges connect to certain pairs of neighboring points.

In general, there are two traditional branches in segmenting point clouds. The first involves purely mathematical models and geometric reasoning techniques like region growing or model fitting. The second involves the extraction of 3D features using feature descriptors and object category classification using machine learning techniques. The first type of methods offers faster computation speeds but it only works well in simple scenarios. Hence, the second type of methods is more commonly used in practice and usually performs better.

Considering the second type of method, segmentation is usually formulated as a pointwise classification problem. Each point is first described by a feature descriptor such as FPFH or SHOT (see Sects. 2.4.2.2 and 2.4.2.3), which rely on handcrafted features and the local geometric properties of points. Then, the extracted features are concatenated to feature vectors and fed into a classifier such as support vector machine (SVM) [27] and random forest (RF) [9]. Representative works include [20, 24]. An overview of [24] for the semantic labeling of 3D point clouds is shown in Fig. 2.24. The first step is a standard pointwise classification process, as illustrated above. The second step describes a method for smoothing the initial labeling by structured regularization. This second part is not discussed here, as it is not the focus of this book.

2.6 Registration

In this section, we discuss some of the traditional methods for point cloud registration, which include the famous iterative closest point algorithm and its variants. Further, some global registration methods are presented.

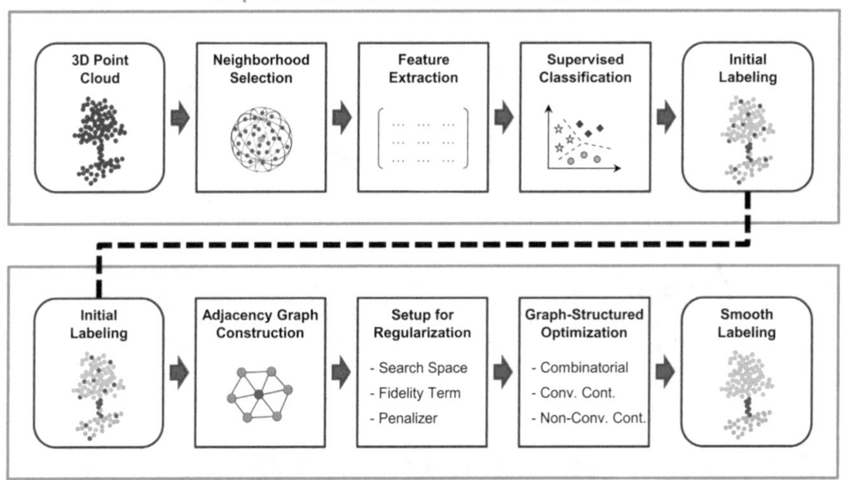

Fig. 2.24 A traditional Semantic Segmentation Pipeline. Reproduced with permission [24]. Copyright © 2010, Elsevier B.V

2.6.1 Iterative Closest Point (ICP)

The classical iterative closest point (ICP) [6] algorithm alternates between two steps: finding point correspondences and estimating the transformation that minimizes the Euclidean distance between matching points. For finding point correspondences, a nearest neighbor search is used. The ICP algorithm is very basic and relies purely on the 3D coordinates of points. Later, we will discuss some methods that rely on some form of point features (handcrafted or learned) for point correspondence.

The ICP algorithm works as follows. Let $A = \{a_i\}$ and $B = \{b_i\}$ be the two sets of point clouds to be registered. We are interested in the transformation T that best aligns the two point clouds. T is a rigid transformation consisting of 3D rotation and translation. Usually, an initial alignment T_0 is obtained using a global alignment algorithm. Otherwise, T_0 is considered an identity transformation. In the first iteration, T is set to T_0 and point cloud B is transformed using T. Then, for every point in point cloud A, its nearest point in the transformed B is found. For example, if every point $T \cdot b_i$ matches with point m_i from point cloud A, we then have an ordered pair of point corresponding points, $(m_i, T \cdot b_i)$. These point correspondences are then used to find the optimal transformation that minimizes the Euclidean distance between corresponding points. It is formulated as

$$T = \arg\min_{T} \left\{ \sum_{i} \|T \cdot b_i - m_i\|^2 \right\}. \tag{2.52}$$

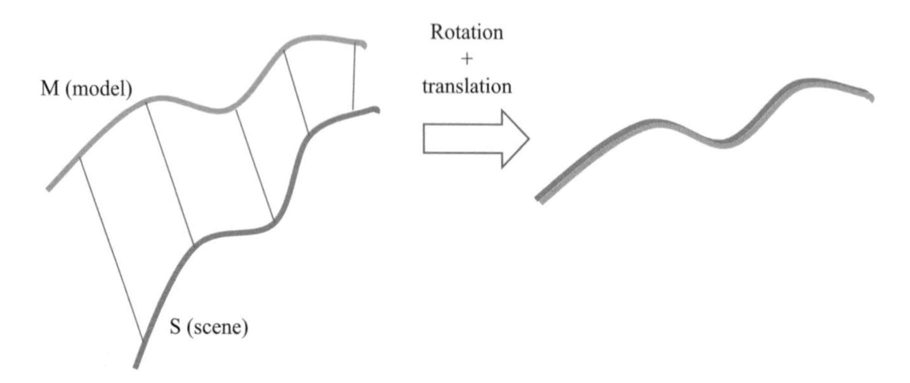

Fig. 2.25 ICP algorithm

The optimal T is found using singular value decomposition (SVD) or a least squares technique. These steps are repeated for a fixed number of iterations or until convergence. There are two main drawbacks of the ICP algorithm. First, it assumes that there is a one-to-one point correspondence between the two point clouds. However, in practice, there may only be a small partial overlap between the two point clouds. Therefore, it is common practice to set a threshold distance d_{max} between two corresponding points, above which the pair of corresponding points is not reliable. Then, only the set of point correspondences with a distance between corresponding points of less than d_{max} is used to estimate the transformation. The second drawback of ICP is that, when the initial alignment is far from the optimal solution, the algorithm can get trapped in a local minimum. In such cases, it is preferable to conduct a prior global registration using a different algorithm, after which ICP can be used to obtain a tighter alignment.

A simple illustration of the ICP algorithm for the alignment of two curves ("model" and "scene") is shown in Fig. 2.25.

2.6.2 Point-to-Plane ICP

Point-to-plane ICP [10] incorporates surface normal information in order to improve the performance of the ICP algorithm. Instead of minimizing the pointwise Euclidean distance (error term), it minimizes the projection of the error term onto the subspace spanned by the surface normal. First, the surface normal of every point in point cloud A is found. If η_i is the surface normal for point m_i, then Eq. 2.52 is modified as follows:

$$T = \arg\min_{T} \left\{ \sum_{i} \|\eta_i \cdot (T \cdot b_i - m_i)\|^2 \right\}. \tag{2.53}$$

2.6.3 Generalized ICP

Generalized ICP [35] replaces the cost function of the original ICP (Eq. 2.52) with a probabilistic model. The steps to find correspondence using a nearest neighbor search are the same as that for ICP. For example, let $(a_i, b_i)_{i=1,\cdots,N}$ be the set of point correspondences found. It is assumed that there exists a set of points $\hat{A} = \{\hat{a_i}\}$ and $\hat{B} = \{\hat{b_i}\}$ that generate the points in point clouds A and B, respectively. The point samples are assumed to be drawn from a normal distribution as $a_i \sim N(\hat{a_i}, C_i^A)$ and $b_i \sim N(\hat{b_i}, C_i^B)$, where C_i^A and C_i^B are the associated covariance matrices. For the correct transformation T^*, the presence of perfect correspondence gives the relation

$$\hat{b_i} = T^* \hat{a_i}. \tag{2.54}$$

Let $d_i^{(T)} = b_i - T a_i$ for any transformation T. Then, $d_i^{(T^*)}$ is governed by the following distribution:

$$\begin{aligned} d_i^{(T^*)} &\sim N(\hat{b_i} - (T^*)\hat{a_i}, C_i^B + (T^*) C_i^A (T^*)^T) \\ &= N(0, C_i^B + (T^*) C_i^A (T^*)^T). \end{aligned} \tag{2.55}$$

This is reduced to a zero mean Gaussian by using Eq. 2.54. Subsequently, maximum likelihood estimation (MLE) is used to iteratively solve for T:

$$\begin{aligned} T &= \arg\max_T \prod_i p(d_i^{(T)}) \\ &= \arg\max_T \sum_i \log(p(d_i^{(T)})). \end{aligned} \tag{2.56}$$

This simplifies to

$$T = \arg\min_T \sum_i d_i^{(T)^T} (C_i^B + T C_i^A T^T)^{-1} d_i^{(T)}. \tag{2.57}$$

Once, T is found using Eq. 2.57, the algorithm repeats. The original ICP algorithm is a special case of the generalized ICP case, where

$$\begin{aligned} C_i^A &= 0 \\ C_i^B &= I, \end{aligned} \tag{2.58}$$

which reduces Eq. 2.57 to

$$T = \arg\min_T \sum_i d_i^{(T)^T} d_i^{(T)} = \arg\min_T \sum_i \|d_i^{(T)}\|^2. \tag{2.59}$$

Similarly, the point-to-plane ICP can be thought of as finding transformation T, such that

$$T = \arg\min_T \left\{ \sum_i \| P_i d_i \|^2 \right\}, \tag{2.60}$$

where P_i is the projection onto the surface normal of b_i. Using the property of an orthogonal projection matrix, $P_i = P_i^2 = P_i^T$. Then, Eq. 2.60 can be equated to

$$T = \arg\min_T \left\{ \sum_i d_i^T P_i d_i \right\}, \tag{2.61}$$

Comparing this to Eq. 2.57, the covariance matrices for point-to-plane ICP are given by

$$\begin{aligned} C_i^A &= 0 \\ C_i^B &= P_i^{-1}. \end{aligned} \tag{2.62}$$

The advantage of generalized ICP is that it allows any set of covariance matrices $\{C_i^A\}$ and $\{C_i^B\}$ to be selected. A direct application of generalized ICP is the plane-to-plane ICP, which considers surface normal information from both point clouds.

2.6.4 Global Registration

The ICP algorithm fails to achieve good alignment in the absence of good initialization; that is, only locally optimal solutions are guaranteed by ICP. There are several methods that offer global registration. One popular method is globally optimal ICP or Go-ICP [43]. It is based on the Branch-n-Bound (BnB) algorithm that searches the entire SE(3) space, which represents all possible transformations. The error term minimization of Go-ICP is similar to that of ICP (Eq. 2.52). Another method is fast global registration (FGR) [45]. FGR facilitates the global registration of multiple partially overlapping 3D surfaces. Teaser [42] proposed a certifiable algorithm that registers two point clouds with a large number of outlier correspondences. Furthermore, a class of algorithms have been developed that couple traditional feature descriptors with RANSAC for robust registration. Detailed discussion on these methods is beyond the scope of this book. We encourage the reader to refer to the original papers for more information.

References

1. Altman, N.S.: An introduction to kernel and nearest-neighbor nonparametric regression. Am. Stat. **46**(3), 175–185 (1992)
2. Bailey, T., Durrant-Whyte, H.: Simultaneous localization and mapping (slam): part II. IEEE Robot. Autom. Mag. **13**(3), 108–117 (2006)
3. Bentley, J.L.: Multidimensional binary search trees used for associative searching. Commun. ACM **18**(9), 509–517 (1975)
4. Bentley, J.L.: Survey of techniques for fixed radius near neighbor searching. Tech. rep., Stanford Linear Accelerator Center, Calif. (USA) (1975)
5. Bentley, J.L., Stanat, D.F., Williams Jr., E.H.: The complexity of finding fixed-radius near neighbors. Inform. Process. Lett **6**(6), 209–212 (1977)
6. Besl, P.J., McKay, N.D.: Method for registration of 3-d shapes. In: Sensor Fusion IV: Control Paradigms and Data Structures, vol. 1611, pp. 586–606. International Society for Optics and Photonics (1992)
7. Brown, M., Lowe, D.G.: Automatic panoramic image stitching using invariant features. Int. J. Comput. Vis. **74**(1), 59–73 (2007)
8. Btyner: K-d tree (2006). https://commons.wikimedia.org/wiki/File:3dtree.png. Accessed 16 Aug 2021
9. Chehata, N., Guo, L., Mallet, C.: Airborne lidar feature selection for urban classification using random forests. In: Laserscanning (2009)
10. Chen, Y., Medioni, G.: Object modelling by registration of multiple range images. Image Vis. Comput. **10**(3), 145–155 (1992)
11. Cormen, T.H., Leiserson, C.E., Rivest, R.L., Stein, C.: Introduction to Algorithms. MIT Press, Cambridge (2009)
12. Dcoetzee: Binary Search Tree (2005). https://commons.wikimedia.org/wiki/File:Binary_search_tree.svg. Accessed 16 Aug 2021
13. Deng, H., Birdal, T., Ilic, S.: PPFNet: Global context aware local features for robust 3d point matching. In: Proceedings of the IEEE Conference on Computer Vision and Pattern Recognition, pp. 195–205 (2018)
14. Duda, R.O., Hart, P.E.: Use of the Hough transformation to detect lines and curves in pictures. Commun. ACM **15**(1), 11–15 (1972)
15. Durrant-Whyte, H., Bailey, T.: Simultaneous localization and mapping: part I. IEEE Robot. Autom. Mag. **13**(2), 99–110 (2006)
16. Eldar, Y., Lindenbaum, M., Porat, M., Zeevi, Y.Y.: The farthest point strategy for progressive image sampling. IEEE Trans. Image Process. **6**(9), 1305–1315 (1997)
17. Finkel, R.A., Bentley, J.L.: Quad trees a data structure for retrieval on composite keys. Acta Inform. **4**(1), 1–9 (1974)
18. Fischler, M.A., Bolles, R.C.: Random sample consensus: a paradigm for model fitting with applications to image analysis and automated cartography. Commun. ACM **24**(6), 381–395 (1981)
19. Fix, E.: Discriminatory Analysis: Nonparametric Discrimination, Consistency Properties, vol. 1. USAF school of Aviation Medicine (1985)
20. Hackel, T., Wegner, J.D., Schindler, K.: Fast semantic segmentation of 3d point clouds with strongly varying density. ISPRS Ann. Photogramm. Remote Sens. Spatial Inform. Sci. **3**, 177–184 (2016)
21. Harris, C.G., Stephens, M., et al.: A combined corner and edge detector. In: Alvey Vision Conference, vol. 15, pp. 10–5244. Citeseer (1988)
22. Katsavounidis, I., Kuo, C.C.J., Zhang, Z.: A new initialization technique for generalized Lloyd iteration. IEEE Signal Process. Lett. **1**(10), 144–146 (1994)
23. Knuth, D.E.: The Art of Computer Programming, vol. 3. Pearson Education (1997)

24. Landrieu, L., Raguet, H., Vallet, B., Mallet, C., Weinmann, M.: A structured regularization framework for spatially smoothing semantic labelings of 3d point clouds. ISPRS J. Photogramm. Remote Sens. **132**, 102–118 (2017)
25. Leon, S.J., Bica, I., Hohn, T.: Linear Algebra with Applications, vol. 6. Prentice Hall, Upper Saddle River (1998)
26. Lowe, D.G.: Object recognition from local scale-invariant features. In: Proceedings of the Seventh Ieee International Conference on Computer Vision, vol. 2, pp. 1150–1157. IEEE, Piscataway (1999)
27. Mallet, C., Bretar, F., Roux, M., Soergel, U., Heipke, C.: Relevance assessment of full-waveform lidar data for urban area classification. ISPRS J. Photogrammetry Remote Sensing **66**(6), S71–S84 (2011)
28. Meagher, D.: Geometric modeling using octree encoding. Comput. Graph. Image Process. **19**(2), 129–147 (1982)
29. Moenning, C., Dodgson, N.A.: Fast marching farthest point sampling. Tech. rep., University of Cambridge, Computer Laboratory (2003)
30. Msm: Noisydata (2007). https://commons.wikimedia.org/wiki/File:Line_with_outliers.svg. Accessed 18 Aug 2021
31. Msm: Ransac (2007). https://commons.wikimedia.org/wiki/File:Fitted_line.svg. Accessed 18 Aug 2021
32. Nguyen, A., Le, B.: 3d point cloud segmentation: a survey. In: 2013 6th IEEE Conference on Robotics, Automation and Mechatronics (RAM), pp. 225–230. IEEE, Piscataway (2013)
33. Rusinkiewicz, S., Levoy, M.: Efficient variants of the ICP algorithm. In: Proceedings Third International Conference on 3-D Digital Imaging and Modeling, pp. 145–152. IEEE, Piscataway (2001)
34. Rusu, R.B., Blodow, N., Beetz, M.: Fast point feature histograms (FPFH) for 3d registration. In: 2009 IEEE International Conference on Robotics and Automation, pp. 3212–3217. IEEE, Piscataway (2009)
35. Segal, A., Haehnel, D., Thrun, S.: Generalized-ICP. In: Robotics: Science and Systems, vol. 2, p. 435. Seattle (2009)
36. Shi, J., et al.: Good features to track. In: 1994 Proceedings of IEEE Conference on Computer Vision and Pattern Recognition, pp. 593–600. IEEE, Piscataway (1994)
37. Sipiran, I., Bustos, B.: Harris 3d: a robust extension of the Harris operator for interest point detection on 3d meshes. Vis. Comput. **27**(11), 963–976 (2011)
38. Smith, S.M., Brady, J.M.: Susan—a new approach to low level image processing. Int. J. Comput. Vis. **23**(1), 45–78 (1997)
39. Tombari, F., Salti, S., Di Stefano, L.: Unique signatures of histograms for local surface description. In: European Conference on Computer Vision, pp. 356–369. Springer, Berlin (2010)
40. WhiteTimberwolf, P.v.N.: Octree (2010). https://commons.wikimedia.org/wiki/File:Octree2.svg. Accessed 16 Aug 2021
41. Wold, S., Esbensen, K., Geladi, P.: Principal component analysis. Chemom. Intell. Lab. Syst. **2**(1–3), 37–52 (1987)
42. Yang, H., Shi, J., Carlone, L.: Teaser: Fast and certifiable point cloud registration. IEEE Trans. Robot. **37**(2), 314–333 (2020)
43. Yang, J., Li, H., Campbell, D., Jia, Y.: Go-ICP: A globally optimal solution to 3d ICP point-set registration. IEEE Trans. Pattern Anal. Mach. Intell. **38**(11), 2241–2254 (2015)
44. Zhong, Y.: Intrinsic shape signatures: A shape descriptor for 3d object recognition. In: 2009 IEEE 12th International Conference on Computer Vision Workshops, ICCV Workshops, pp. 689–696. IEEE, Piscataway (2009)
45. Zhou, Q.Y., Park, J., Koltun, V.: Fast global registration. In: European Conference on Computer Vision, pp. 766–782. Springer, Berlin (2016)
46. Zhou, Q.Y., Park, J., Koltun, V.: Open3d: A modern library for 3d data processing (2018). arXiv preprint arXiv:1801.09847

Chapter 3
Deep Learning-Based Point Cloud Analysis

Abstract Deep learning has achieved impressive performance improvements over traditional methods for almost all vision tasks. Point cloud processing is no exception. Since 2017, researchers have become inclined to train end-to-end networks for tasks like point cloud classification, semantic segmentation, and object detection. More recently, other tasks like registration and odometry have also been solved using Deep learning. These newer data-driven methods provide some benefits over traditional methods that rely on handcrafted features. Nevertheless, many traditional methods are still in practice due to their simplicity and speed, and they form the basis of newer methods. In this chapter, we discuss some Deep learning-based methods for point cloud processing. This subset of methods has had a huge impact in this field and is representative of current research progress in computer vision. The Deep learning methods for point cloud classification, semantic segmentation, and registration tasks are discussed. We explore several papers, with a focus on the proposed methods and associated details, while the experimental details are limited to performance evaluations on benchmark datasets. Other analyses such as ablation studies and miscellaneous details from the papers are omitted.

3.1 Introduction

Neural networks and Deep learning have had a huge impact on the field of computer vision. The success of AlexNet [14] on the ImageNet Large-Scale Visual Recognition Challenge (ILSVRC), a benchmark test for object category classification and detection tasks, shifted the focus of researchers to develop similar networks for the image classification task. This led to a large number of convolutional neural networks (CNNs) being proposed for a range of vision tasks like image classification, semantic segmentation, object detection, tracking, video processing, and so on.

Point cloud processing was also on the rise, with growing research on autonomous driving, robotic vision systems, applications in computer graphics, and so on. Until 2017, most of the methods and algorithms that had been developed for these purposes were based on traditional handcrafted features. We have seen

S. Liu et al., *3D Point Cloud Analysis*,
https://doi.org/10.1007/978-3-030-89180-0_3

some of these noteworthy methods in the previous chapter. These methods rely heavily on the local 3D geometric properties of points. As point clouds are an unstructured form of data, it was nontrivial to extend CNNs to 3D point clouds directly, even though they had proven highly effective on 2D image data. Some intermediate research then emphasized the process of converting point clouds to regular forms like voxel grids or projecting them to multi-view images. This allowed the power of CNNs and Deep learning to be harnessed based on the newly structured data. However, these methods possessed several bottlenecks, such as the large computation time for conversion, sparsity of voxel grids, and loss of information during sampling.

In 2017, a deep network, PointNet [18], was applied to point sets directly for the first time, without any preprocessing or conversion to other forms. Soon, a follow-up method called PointNet++ [19] was proposed that resolved some of the drawbacks of PointNet. This marked the beginning of a new era in 3D point cloud processing. PointNet and PointNet++ formed the basis of more, deeper networks for object classification, registration, semantic segmentation, and detection tasks.

Today, there are several notable deep-learning-based methods. They try to solve different problems associated with large-scale point cloud processing. Much inspiration has been taken from 2D vision research, graph signal processing, and, very recently, attention and transformers. In this chapter, we present a selective review of some of the most impactful methods. The goal is to equip the reader with some knowledge of the most popular methods that summarize a larger group of methods and the overall research direction. The methods have been divided into two sections—one for classification and segmentation and one for point cloud registration.

3.2 Classification and Segmentation

In this section, we will go over some of the most impactful works in point cloud object classification, part segmentation, and semantic segmentation tasks: PointNet, PointNet++, DGCNN, PointCNN, PointSIFT, Point Transformer, and RandLA-Net.

3.2.1 PointNet

PointNet [18] was the first work to employ Deep learning directly for processing 3D point clouds. Prior to PointNet, 3D point clouds were transformed to regular representations such as voxels or multiple 2D images before processing.

The authors of PointNet considered that, in order to design a deep network that works for point clouds, some desired properties must be taken into account. These properties include the unordered nature of points, interactions of points in a local region, and invariance to certain geometric transformations. Point clouds being an unordered set, the network has to learn to be invariant to the order in which the points are fed to the network. In short, for a point cloud with N points, the network should be invariant to $N!$ permutations of points. Besides, the points are not isolated entities; points in a neighborhood define a local structure. Hence, the model needs to capture these interactions in a local region. Finally, it is desirable for the model to learn to output the same label or semantic category under rotation, translation, and any affine transformation of the point cloud. PointNet was designed to maintain these properties.

To solve the order invariance problem, PointNet uses a symmetric set function in the form of max pooling. The max pooling operator does not depend on the order of operands (here, the points) and hence is suitable for order invariance. The authors also discuss some other strategies such as sorting input points in a canonical order, treating point clouds as sequential data, and using recurrent neural networks (RNNs). However, these techniques are experimentally less effective than symmetric functions. PointNet uses a multilayer perceptron (MLP), which is a universal function approximator, to approximate the set function of assigning a class label or semantic category to a point. The network is mathematically formulated as

$$f(x_1, x_2, \ldots, x_n) \approx g(h(x_1), h(x_2), \ldots, h(x_n)), \tag{3.1}$$

where f is the underlying function of the points (a class category for the classification task or a per-point label for the segmentation task), and $x_i s$ are the input points. PointNet approximates h using a MLP, and a composition of a single variable functions and max pooling is used to find g. The detailed network is depicted in Fig. 3.1.

The network takes n points, represented by their 3D coordinates, and first applies an input transform using a T-Net. The purpose of this is to ensure the input is invariant to geometric transformations. The T-Net is like a mini-PointNet network that learns a 3×3 affine transformation matrix. A sequence of pointwise MLPs transform points to higher dimensional feature space. Then, a separate feature

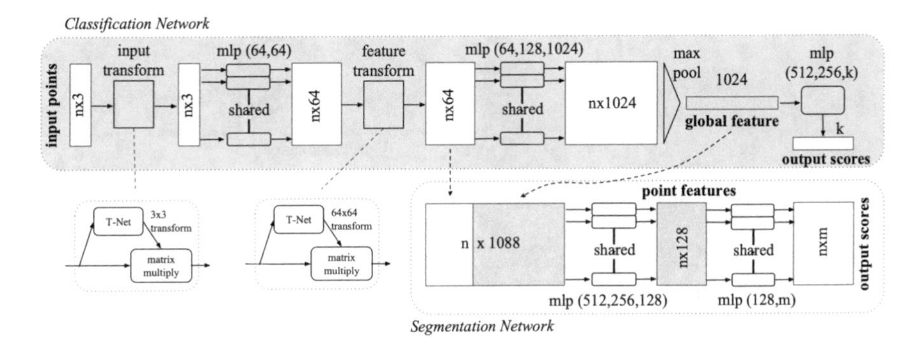

Fig. 3.1 PointNet architecture. Reproduced with permission [18]. Copyright © 2017, IEEE

transform is applied with the same purpose, to make the features invariant to transformations. The point features are aggregated using max pooling to obtain a 1024-dimensional global feature vector. For the classification task, this feature vector is further fed to a MLP classifier, which outputs a k-dimensional probability vector for k classes. For the segmentation task, the global feature is fused with the pointwise features to further learn the output labels for each point. The concatenation of local and global features enables the network to leverage both local geometry and global semantics.

The potential of PointNet has been experimentally highlighted. For object classification tasks, PointNet achieves an overall accuracy of 89.2% on the ModelNet40 dataset [26], which is greater than all previously developed voxel-based methods. For the part segmentation task, PointNet achieves a mean Intersection-over-Union (IoU) of 83.7% on the ShapeNet dataset [30]. The mean IoU and overall accuracy are, respectively, 47.71% and 78.62% for semantic segmentation on the S3DIS dataset [3]. Analysis has revealed that the model has learned to summarize shapes using a smaller set of representative points. Furthermore, it is robust to small perturbations of input points, as well as to outliers and missing points.

3.2.2 PointNet++

PointNet shows impressive performance for point cloud processing tasks like classification and segmentation. However, it does not capture information about the local context of points at different scales. In a follow-up work, termed PointNet++ [19], the researchers proposed a hierarchical feature learning framework to resolve some limitations of PointNet. The hierarchical learning process is achieved by a series of set abstraction levels. Each set abstraction level consists of a sampling layer, grouping layer, and PointNet layer. The PointNet++ architecture is shown in Fig. 3.2.

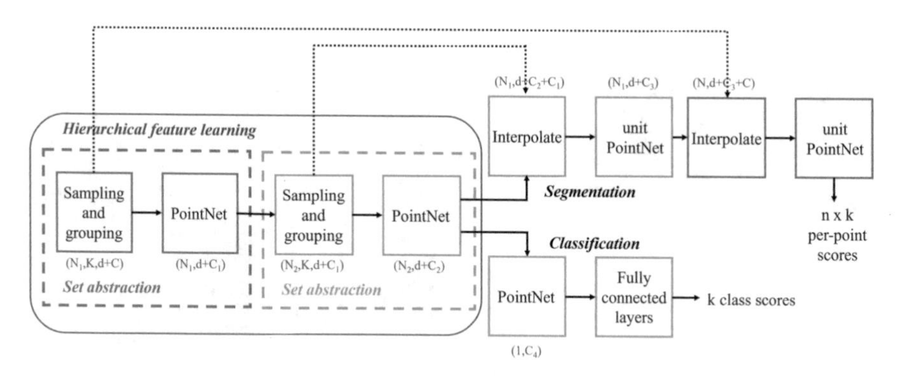

Fig. 3.2 PointNet++ architecture

- **Sampling**—In the sampling layer, a subset of m points $\{x_{i1}, x_{i2}, \ldots, x_{im}\}$ is sampled from the input n points. The iterative farthest point sampling (FPS) technique is used here, which provides uniform coverage of the entire point cloud. These sampled m points form the set of centroids for the grouping layer.
- **Grouping**—The grouping layer takes the input point set of size $N \times (d + C)$, where d corresponds to the dimension of the coordinates and C is the feature dimension, and the coordinates of the centroids of dimension $N' \times d$ from the sampling step. All the points lying inside a sphere of a certain radius around each centroid are collected. The output grouped point set is of size $N' \times K \times (d + C)$, where K is the number of points lying within the sphere. K varies for every point depending on the density of points in the local region.
- **PointNet**—For each set of grouped points, the point coordinates are first translated to a local system centered at the centroid. Then, the PointNet operation is performed in the local region, in the manner discussed in Sect. 2.1.1. The features of all K points are aggregated using a local max pooling operation. The output is of size $N' \times (d + C)$

Point cloud feature learning is affected by the non-uniformity of point densities in different regions. To tackle the problem of varying point densities, PointNet++ introduces density-adaptive PointNet layers that combine features at different scales in the presence of different sampling densities. This is achieved using multiscale grouping (MSG) and multiresolution grouping (MRG):

- **Multiscale grouping** (**MSG**)—A natural way to solve the density problem is to combine information from different scales. As the scale increases (greater radius of sphere), the number of points included in the grouping layer is increased. By fusing the features from different scales, a multiscale feature representation is formed.
- **Multiresolution grouping** (**MRG**)-Point features are formed from a concatenation of two vectors. One vector is obtained from the lower set abstraction level by summarizing the features in the subregions. The other vector is the result of the PointNet operation in the current set abstraction level. The weight of each vector is adjusted based on the point densities in the present and previous set abstraction levels.

MSG has a higher computation cost than MRG, because the PointNet layer needs to be applied multiple times for every point in MSG, but only once for a single scale at a given set abstraction level in MRG.

For the segmentation task, it is necessary to have the features for every point. For this, a layer-wise interpolation technique is used, which helps propagate the point features hierarchically. The features of the N_l points at the l-th set abstraction level are used to obtain the features of the N_{l-1} points at the set abstraction layer $l - 1$. Note that $N_l \leq N_{l-1}$. To interpolate the features for a point, the features of the k-nearest points are used. The k features are weighted using the inverse distance to the point. Three nearest neighbors are considered. The interpolated features are then passed through a unit PointNet layer, which includes shared fully connected layers. For the classification task, all features at the deepest set abstraction layer are max pooled to obtain the global feature vector for the object. Thereafter, a MLP classifier is used to obtain the output scores. Feature interpolation is not required for classification.

The performance of PointNet++ is better than PointNet. It achieves an overall accuracy of 90.7% on the object classification task when evaluated on the ModelNet40 dataset. When point normal information is combined with point coordinates, the overall accuracy further improves to 91.9%. Furthermore, the ability of PointNet++ for point cloud classification in non-Euclidean metric space and its robustness to varying sampling densities have been highlighted experimentally.

3.2.3 Dynamic Graph CNN

Dynamic Graph CNN (DGCNN) [25] is a novel method that uses EdgeConv operators to help capture the local neighborhood information of points. EdgeConv acts on graphs that are computed from a point cloud in every layer of the network. Multiple layers involving EdgeConv are cascaded to learn global shape information. The proposed DGCNN network can learn to semantically group points through dynamic graph updates from layer to layer, while EdgeConv can be integrated into multiple existing pipelines for point cloud processing.

The EdgeConv operation functions as follows. The main idea is similar to that of graph neural networks. A local neighborhood graph is constructed from a point cloud, after which a convolution-like operator is applied to the edges. For a given layer in the network, let $\mathrm{X} = \{x_1, x_2, \cdots, x_n\} \in \mathbb{R}^F$ be the set of n points with dimension F. In the first layer, F is usually equal to 3, which denotes the 3D coordinates of points; however, additional information such as color and surface normal may also be included. Then, a directed graph $G = (V, E)$ is computed, wherein the 3D points represent the vertices, $V = 1, 2, \cdots, n$ and $E \subseteq V \times V$ are the edges. A simple example of such a graph is one with edges to the k-nearest neighbors of every point in F dimensional space. The feature of every edge is calculated as $e_{ij} = h_\theta(x_i, x_j)$, where $h_\theta : \mathbb{R}^F \times \mathbb{R}^F \rightarrow \mathbb{R}^{F'}$ is a nonlinear function with trainable parameters θ. Further, a channelwise symmetric function $\square$ such as max or summation is applied on the edge features of all edges originating from every vertex. For the i-th vertex, the output of EdgeConv is given by

$$x_i' = \mathop{\square}_{j:(i,j)\in E} h_\theta(x_i, x_j). \tag{3.2}$$

This outputs n points with dimension F'. The choice of h and $\square$ is crucial in the design of EdgeConv, and it directly affects the performance of DGCNN. While several potential choices are discussed, the one that works best combines global shape structure and local neighborhood information, as given by Eq. 3.3:

$$h_\theta(x_i, x_j) = h_\theta(x_i, x_j - x_i). \tag{3.3}$$

Here, x_i provides global coordinate information about the i-th point, while $x_j - x_i$ considers the local neighborhood structure. Then, the edge features are calculated as

$$e'_{ijm} = \text{ReLU}(\theta_m \cdot (x_j - x_i) + \phi_m \cdot x_i). \tag{3.4}$$

This process is implemented using a shared MLP. The parameters in Eq. 3.4 are given by $\Theta = (\theta_1, \cdots, \theta_M, \phi_1, \cdots, \phi_M)$ where M is the number of filters used. Max is used as the symmetric function. Therefore, the output of EdgeConv is

$$x'_{im} = \max_{j:(i,j)\in E} e'_{ijm}. \tag{3.5}$$

The EdgeConv operation is illustrated in Fig. 3.3.

It has been empirically shown that the graph should be updated in every layer, unlike the static graphs used in most methods. Accordingly, at each layer, the graph is recomputed as follows. A feature distance matrix is calculated for all points. The feature dimension is F, as discussed previously. There are edges between points (vertices) and all the k neighboring points in terms of feature distance in the new graph $G^{(l)} = (V^{(l)}, E^{(l)})$ at layer l. Hence, there are k directed edges for each vertex. EdgeConv has properties like permutation invariance and partial translation invariance.

The DGCNN architecture is illustrated in Fig. 3.4. It consists of stacked layers that conduct EdgeConv operations on the constructed graphs. The classification and

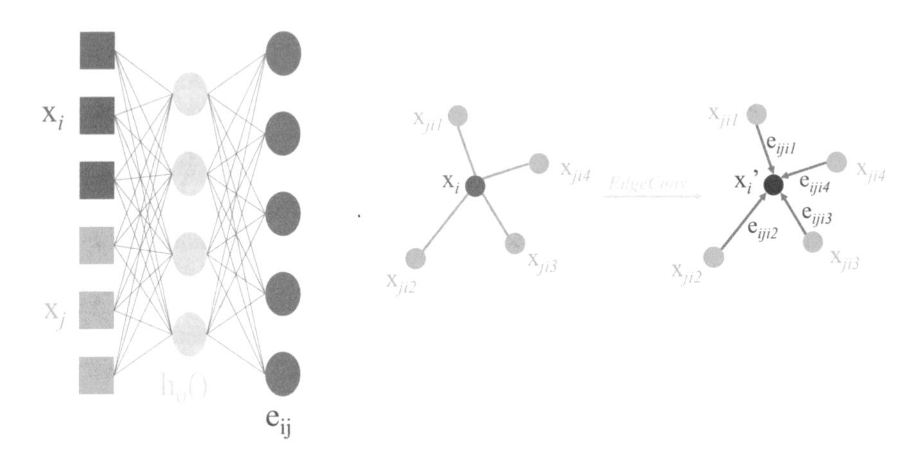

Fig. 3.3 EdgeConv operation in DGCNN

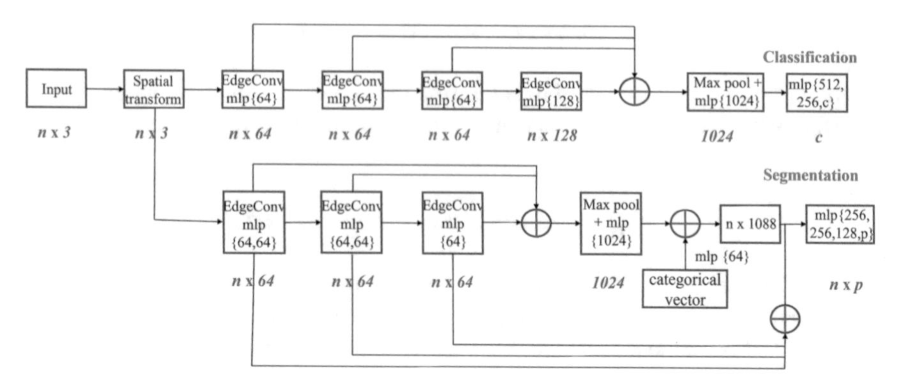

Fig. 3.4 DGCNN architecture

segmentation network are shown. Unlike PointNet++, the point cloud downsampling operation is absent in DGCNN, and in every layer, the same number of points is used.

DGCNN achieves an impressive accuracy of 93.5% for point cloud classification on the ModelNet40 dataset. The mean IoU for the ShapeNet part segmentation dataset is 85.2%, while the overall accuracy for semantic segmentation on the S3DIS dataset is 84.1%.

3.2.4 PointCNN

Inspired by the successes of using CNNs for processing image data, PointCNN [15] proposes a method for learning a χ-Transformation that can transform input points to a feature representation on which a convolution operation can be applied. The goal of the χ-Transformation is to learn an order invariant $K \times K$ matrix for K input points using a MLP, $\chi = MLP(p_1, p_2, \cdots, p_K)$. Under the permutation invariant case, the output of the convolution operation $f = Conv(K, \chi, [p_1, p_2, \cdots, p_K]^T)$ for convolution kernel K will be the same irrespective of the order of input points.

CNNs benefit from hierarchical feature learning where the input feature maps are successively convolved in every layer. After each convolution operation, the spatial resolution of the image is reduced with an increase in the number of channels (or amount of spectral information). A similar hierarchical learning method is adopted by PointCNN for point clouds as follows. The input to a layer of PointCNN is $\mathbb{F}_1 = \{(p_{1,i}, f_{1,i}) : i = 1, 2, \cdots, N_1\}$, where $\{p_{1,i} : p_{1,i} \in \mathbb{R}^{Dim}\}$ is the set of points and $\{f_{1,i} : f_{1,i} \in \mathbb{R}^{C_1}\}$ is the set of features. PointCNN seeks to apply χ-Conv to $\mathbb{F}_1$ to get set $\mathbb{F}_2 = \{(p_{2,i}, f_{2,i}) : i = 1, 2, \cdots, N_2\}$ where $\{p_{2,i} : p_{2,i} \in \mathbb{R}^{Dim}\}$ is the set of points and $\{f_{2,i} : f_{2,i} \in \mathbb{R}^{C_2}\}$ are the output features. Similar to the image case, $N_2 < N_1$, meaning a smaller spatial resolution and $C_2 > C_1$ indicate deeper feature channels.

Next, we review the process of transforming $\mathbb{F}_1$ to $\mathbb{F}_2$ using χ-Conv. Let p be one representative point from the set $\{p_{2,i}\}$ of points in $\mathbb{F}_2$ and f be its feature to be learned. First, a set of K nearest neighbors of p in the set of input points $\{p_{1,i}\}$ is retrieved. This set is termed as $\mathbb{N}$. An unordered set of neighboring points of p and corresponding features is then given by $\mathbb{S} = \{(p_i, f_i) \ : \ p_i \in \mathbb{N}\}$. $\mathbb{S}$ is formed by a $K \times Dim$ matrix **P** of points, $P = (p_1, p_2, \cdots p_K)^T$ and a $K \times C_1$ matrix **F** of features, $F = (f_1, f_2, \cdots f_K)^T$. Let **K** denote the convolution kernels to be learned. The algorithm takes **K**, p, **P**, and **F** and outputs the aggregated feature $\mathbf{F_p}$. The steps are follows.

- The set P is centered at p to obtain P' as P' $\leftarrow$ P - p.
- The points are individually lifted to a higher dimensional space (C_δ dimensions) using MLP as $F_\delta \leftarrow MLP_\delta$(P'). This pointwise MLP operation is similar to PointNet.
- F_δ and F are concatenated to obtain a $K \times (C_\delta + C_1)$ dimensional matrix F_* as $F_* \leftarrow [F_\delta \ F]$
- The $K \times K$ χ-transformation matrix is learned from P' as $\chi \leftarrow MLP$(P'). χ is aware of the order of points in P' and helps to achieve permutation invariance in the next step.
- F_* is weighed and permuted with χ to obtain the order invariant neighboring features F_χ.
- Convolution of F_χ with kernel **K** yields the output F_p as $F_p \leftarrow Conv(K, F_\chi)$.

PointCNN architectures for the classification and semantic segmentation tasks are shown in Fig. 3.5, where architectures (a) and (b) are for the classification task

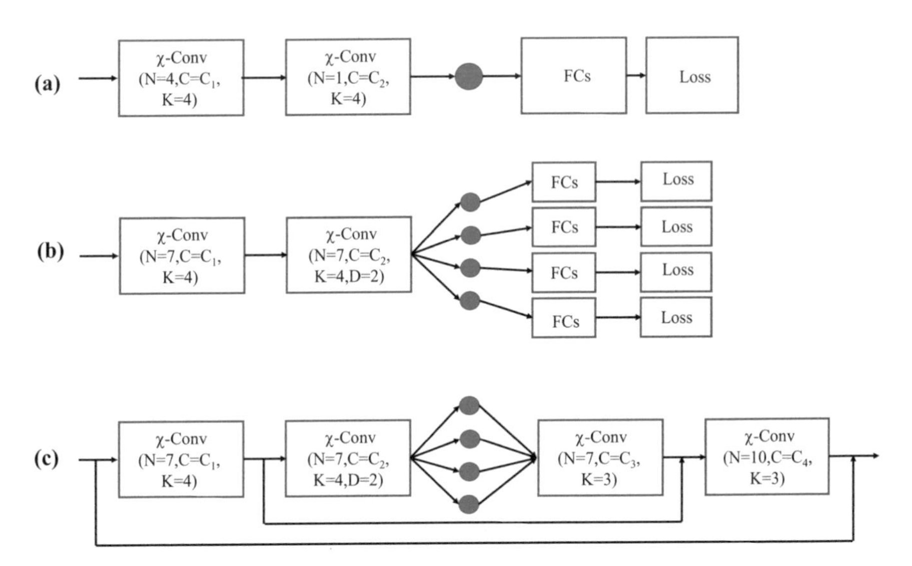

Fig. 3.5 PointCNN architecture. N denotes number of points, C is the feature dimension, K is number of nearest neighbors, and D is dilation rate of χ-Conv

and (c) is for the segmentation task. The classification network consists of multiple χ-Conv layers followed by a fully connected layer for class score prediction. The segmentation network is a U-Net-like network with Conv and DeConv layers. The DeConv operation is similar to χ-Conv, the only difference being that it has more points and fewer channels in the output than the input. DeConv is responsible for dense per-point class predictions.

PointCNN shows impressive results on datasets like ModelNet40, ShapeNet, ScanNet, and S3DIS. It outperforms PointNet and PointNet++ for point cloud classification on the ModelNet40 dataset, achieving an overall accuracy of 92.5%. For the challenging semantic segmentation task, the mean IoU is 65.39% on the S3DIS dataset. Several other experiments including ablation studies and visualization of χ-transformed features have also highlighted the effectiveness of PointCNN.

3.2.5 *PointSIFT*

Scale Invariant Feature Transform (SIFT) [16] is a widely used 2D keypoint detector and descriptor. SIFT is very robust to different image scales and rotations. Inspired by SIFT, PointSIFT [12] designs a scale- and orientation-aware module for 3D point cloud semantic segmentation tasks. However, unlike SIFT, which uses handcrafted features, PointSIFT benefits from feature learning using deep networks. The PointSIFT module can be inserted in PointNet-like architectures to enrich the point features. The main contribution of PointSIFT is the orientation encoding unit that convolves features of nearest neighbors in eight orientations. The set abstraction and feature propagation units are borrowed from PointNet++ to complete the network. Finally, scale awareness is achieved by stacking multiple PointSIFT modules.

The semantic segmentation problem is formulated as follows. An input point cloud P is given which consists of n points having d features, $P = \{p_1, p_2, \cdots, p_n\} \in \mathbb{R}^d$. The set of semantic labels is L. Then, the goal is to learn a mapping function Ψ that assigns a semantic label to each of the n points as follows:

$$\Psi = P \to L^n. \tag{3.6}$$

Next, we study the PointSIFT module in detail. The input to this module is an $n \times d$ matrix of n points having d dimensional features. The output is again a matrix of size $n \times d$, but with new features. The first step is orientation encoding. The authors argue that using an ordered operation like convolution is more effective than using an unordered operation like max pooling. A natural order is induced by ordering the point coordinates. For every point, information from the eight orientations is collected and integrated. This is done in two steps. First, a stacked 8-neighborhood (S8N) search operation collects neighboring points in each of the eight octants formed by partitioning the three coordinates. Next, the features residing in a $2 \times 2 \times 2$ cube are processed using orientation encoding convolution, which convolves the cube along the X, Y, and Z axes successively. The features of neighboring points can be represented as a vector of size $2 \times 2 \times 2 \times d$. Then, the output feature is obtained using the following operations:

$$\begin{aligned} V_x &= \text{ReLU}(Conv(W_x, V)) \in \mathbb{R}_{1x2x2xd} \\ V_{xy} &= \text{ReLU}(Conv(W_y, V_x)) \in \mathbb{R}_{1x1x2xd} \\ V_{xyz} &= \text{ReLU}(Conv(W_z, V_{xy})) \in \mathbb{R}_{1x1x1xd}. \end{aligned} \tag{3.7}$$

Here, $W_x \in \mathbb{R}_{2\times1\times1\times d}$, $W_y \in \mathbb{R}_{1\times2\times1\times d}$, and $W_z \in \mathbb{R}_{1\times1\times2\times d}$ are the weights of convolution in the X, Y, and Z axes, respectively.

Scale awareness is achieved by stacking multiple orientation encoding (OE) units to provide a multiscale representation. Different OE units have different receptive fields. Finally, the outputs of all OE units are concatenated, and another pointwise convolution is performed to obtain the output feature with d-dimensions. The PointSIFT module comprising stacked OE units is shown in Fig. 3.6. The end-to-end optimization process ensures that the network learns to select appropriate scales.

The overall architecture of PointSIFT is illustrated in Fig. 3.7. It consists of an encoder–decoder-style structure for semantic segmentation. The set abstraction step takes an input of $N \times d$, where N is the number of points and d is the feature dimension. The output is N' points with dimension d', and $N > N'$, $d < d'$. In the downsampling step, N' centroids are found using farthest point sampling (FPS). The features of these points of dimension d are learned using shared PointNet. Feature propagation is performed using linear interpolation of nearest neighbor features to upsample and obtain dense representations.

PointSIFT exhibits improved performance compared to PointNet and PointNet++. The overall accuracy of PointSIFT on the S3DIS dataset is 88.72%, while the mean IoU is 70.23%.

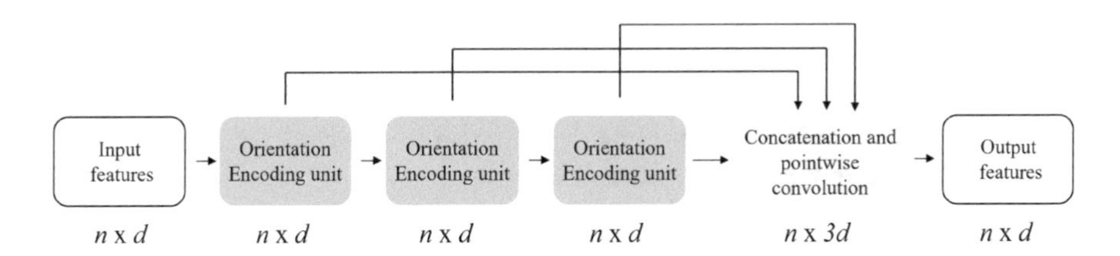

Fig. 3.6 PointSIFT module

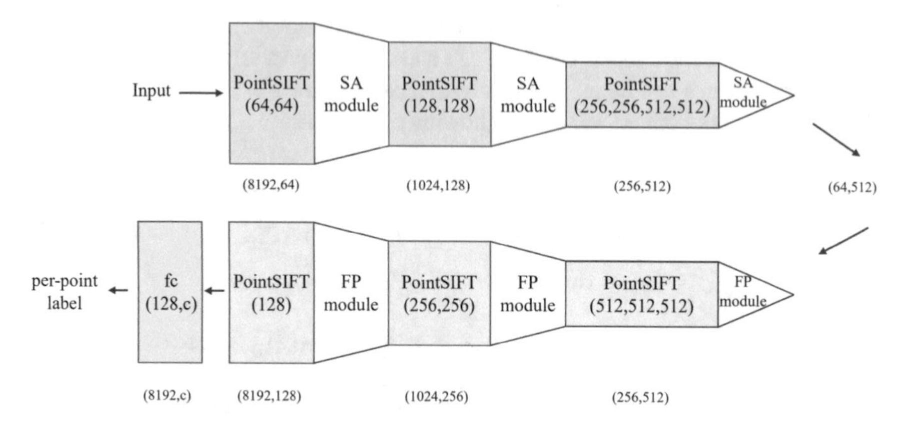

Fig. 3.7 PointSIFT architecture. SA stands for Set Abstraction and FP stands for Feature Propagation

3.2.6 *Point Transformer*

The Transformer model [22] revolutionized the field of natural language processing (NLP). Transformer-based networks have achieved state-of-the-art performance in tasks like language translation and other sequence-to-sequence problems in NLP. Transformer uses a self-attention mechanism to learn contextual information in sequences. Similar methods have subsequently been developed for vision tasks like image classification and object detection. Inspired by these works, Point Transformer [32] incorporates self-attention layers for point cloud processing. The self-attention operator, which is the core of Transformer, is a set operator with positional attributes, thereby being invariant to the order and number of elements on which it is applied. This makes it favorable for processing point clouds, which are unordered sets of points that naturally come with position information in the form of 3D coordinates.

Here, we first review the Point Transformer layer that forms the backbone of this method, and we then discuss the overall architecture. Self-attention operators can be classified as scalar or vector. Point Transformer is designed using vector self-attention. This operation is formulated in Eq. 3.8.

$$y_i = \sum_{x_j \in \chi(i)} \rho(\gamma(\varphi(x_i) - \psi(x_j) + \delta)) \odot (\alpha(x_j) + \delta). \tag{3.8}$$

Here, $\chi(i)$ is the set of k neighboring points of point x_i. To compute the features of a point, the attention operator performs a series of steps as per Eq. 3.8. It first applies a pointwise feature transformation on the features of the input point and its neighboring points using functions φ and ψ, respectively. In attention literature, linear projections or MLPs are used as functions. Point Transformer uses a simple linear projection for φ, ψ, and α. Subsequently, a subtraction operation is conducted as a relation function. This is followed by the mapping function γ, which is a MLP that produces attention vectors. δ represents the position encoding term which helps the attention operator to adapt to local structures. It consists of a learnable function θ modeled using MLP and is given by

$$\delta = \theta(p_i - p_j), \tag{3.9}$$

where p_i and p_j are the point coordinates.

To summarize, the point transformer block takes input points and their features and uses vector attention to produce a new set of point features that depend both on the input features and the distribution of points in 3D space.

The system architecture of the Point Transformer network is illustrated in Fig. 3.8. The network comprises five operations: Point Transformer, transition down, transition up, MLP, and global average pooling. The Point Transformer block is explained in the previous paragraphs, while the other four operations are described below.

The network consists of five stages (or layers). Each stage operates on a downsampled point set from the previous stage. At every stage, the point cloud is downsampled by a factor of 4. The transition down and transition up modules connect two successive stages in the feature encoder and feature decoder, respectively.

Transition Down Module The goal of transition down module is to reduce the number of points from N to $N/4$ while transiting from one stage to the next. A subset of $N/4$ points is sampled from N points using farthest point sampling (FPS). The feature vectors for these $N/4$ points are formed as follows. First, the k nearest neighbors among the input N points are found for every point in the downsampled set. The features of all points in this local set go through a MLP (separately for each point). Finally, the output features are max pooled to obtain the feature for that point.

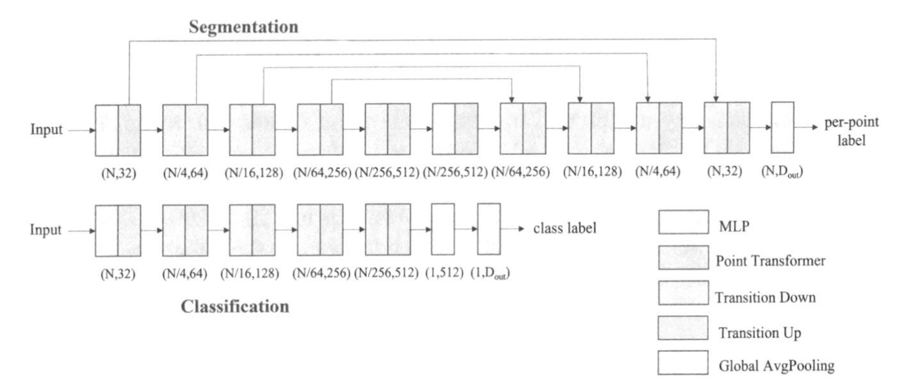

Fig. 3.8 Point Transformer architecture

Transition up Module The transition up module is part of the feature decoder, which increases the resolution of the point cloud. It maps the point features from the downsampled set to its super set, which has more points. This is essential for segmentation, which requires features for every point to make a decision. Each point feature is processed by a mini network comprising a linear layer, batch normalization, and rectified linear unit (ReLU), followed by trilinear interpolation to a higher resolution point set. The interpolated features are summarized using the features of the encoder from the same stage.

Global Average Pooling Module The role of global average pooling is to pool the point features to a single global feature vector for the classification task.

MLP The MLP is first used at the input stage to the first point transformer unit. It takes the input point coordinates and encodes a 32-dimensional feature representation. Next, the MLP is used as a classifier. For the object classification task, the MLP takes the global point cloud feature vector through a series of fully connected layers to output the class scores. For the segmentation task, the output of the last decoder provides pointwise features that are passed through a shared MLP to obtain the per-point prediction scores.

The self-attention mechanism helps Point Transformer achieve better performance than most other methods. For the object classification task, it achieves an overall accuracy of 93.7% and mean class-wise accuracy of 90.6% on the ModelNet40 dataset. In the semantic segmentation task, Point Transformer attains an impressive overall accuracy of 90.8%, a mean accuracy of 76.5%, and a mean IoU of 70.4% when evaluated on Area 5 of the S3DIS dataset.

3.2.7 RandLA-Net

RandLA-Net[11] efficiently performs semantic segmentation for large-scale 3D point clouds by using random point sampling instead of more complex point selection approaches. Recent approaches for directly processing point clouds, such as PointNet/PointNet++ [18, 19], DGCNN [25], PointCNN [15], and PointSIFT [12], achieve impressive results for object recognition and semantic segmentation tasks. However, they are limited to small-scale point clouds. For example, each room of the S3DIS dataset is divided into 1×1 m blocks, each downsampled to 4096 points as a sample. That is, these methods cannot be extended to large-scale point clouds directly, e.g., millions of points and up to 200×200 m. There are three reasons for the incapability of these methods for processing massive numbers of points. First, the point sampling methods adopted currently are computationally expensive or memory inefficient; second, the existing local feature learners rely on expensive kernelization or graph construction; third, these local feature learners are unable to capture complex structures in the large-scale point clouds due to their limited size of receptive fields. The design of RandLA-Net targets at two aspects: the sampling method and the local feature learner.

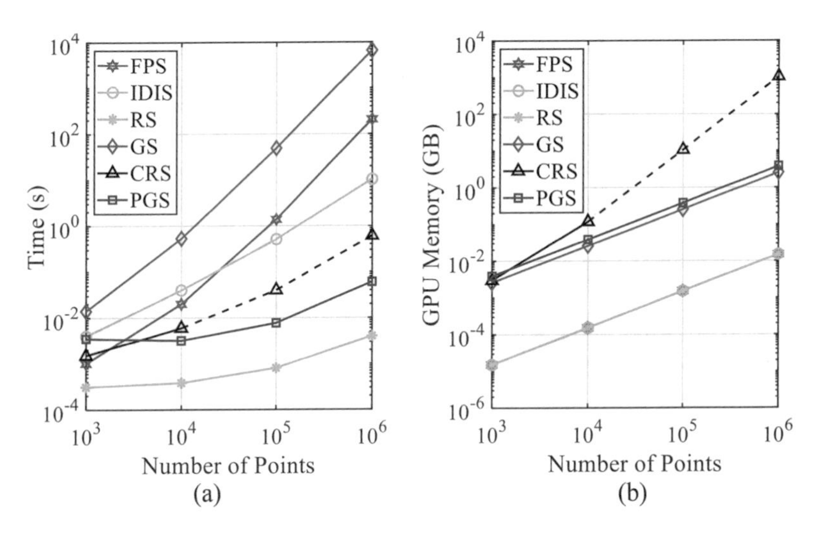

Fig. 3.9 Time and memory consumption of different sampling approaches. Reproduced with permission [11]. Copyright © 2020, IEEE. (**a**) Time consumption (**b**) Memory consumption

Sampling To directly process large-scale point clouds in a single pass, the sampling method should be both memory and computationally efficient so that it can be processed by GPUs. The computation time and GPU memory consumption of existing sampling methods are compared in Fig. 3.9. Farthest point sampling (FPS),inverse density importance sampling (IDIS) [9], and random sampling (RS) are heuristic sampling methods; whereas generator-based sampling (GS) [8], continuous relaxation based sampling (CRS) [1, 29], and policy gradient-based sampling (PGS) [28] are learning-based sampling methods. Details are omitted here. In general, the learning-based methods do not perform as well as the heuristic methods in terms of time or memory cost so far. Specifically, GS is too computationally expensive, CRS's memory cost is high, and PGS is hard to learn. FPS and IDIS are the most frequently used for small-scale point clouds. However, their computation time increases dramatically as the number of points increases, acting as a significant bottleneck to real-time processing. RS is the most suitable sampling method for large-scale point cloud processing, as it is faster and scales more efficiently than existing alternatives. Therefore, this is the method adopted by RandLA-Net.

Local Feature Aggregation Despite the advantages of RS, it is not as accurate as the other sampling methods, as prominent point features may be dropped by chance. To overcome this issue, a new local feature aggregation module that increases the receptive field size in each layer progressively was designed so that complex local structures can be learned effectively. As is shown in Fig. 3.10, each layer comprises a dilated residual block (DRB), which is a stack of multiple local spatial encoding (LocSE) and attentive pooling units with a skip connection. For each point, the module first observes its K nearest neighbors after one LocSE and attentive pooling unit, and then observes the K^2 neighboring points after the second unit.

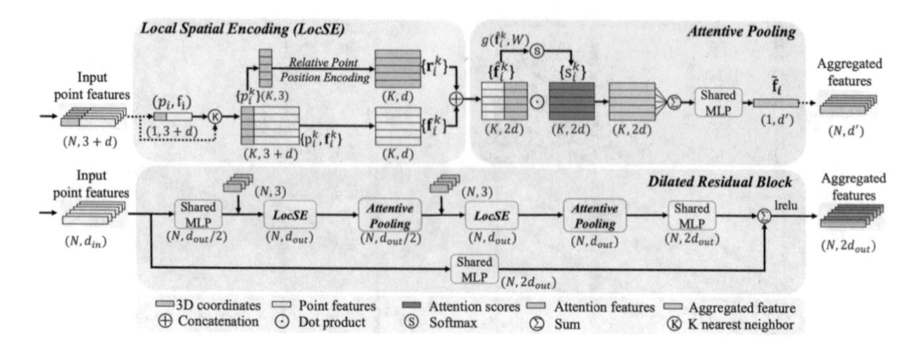

Fig. 3.10 Local feature aggregation module. Reproduced with permission [11]. Copyright © 2020, IEEE

Experimentally, two units are stacked to achieve both efficiency and effectiveness. Given a center point p_i, LocSE first gathers its K nearest neighboring points $\{p_i^1, p_i^2, \cdots, p_i^K\}$ by a simple K-NN algorithm. Then, the relative point position is encoded as:

$$\mathbf{r}_i^k = MLP(p_i \oplus p_i^k \oplus (p_i - p_i^k) \oplus \|p_i - p_i^k\|), \tag{3.10}$$

where p_i and p_i^k are the x-y-z coordinates of the points, and $\oplus$ is the concatenation operation. The encoded relative point position $\mathbf{r}_i^k$ and its corresponding point features $\mathbf{f}_i^k$ are concatenated as an augmented feature $\hat{\mathbf{f}}_i^k$. Instead of using max/mean pooling to the neighboring features like in PointNet/PointNet++ [18, 19], which leads to a lot of information loss, attentive pooling is adopted to learn important local features. A shared function $g()$, a shared MLP followed by *Softmax*, is designed to learn a unique attention score for each feature:

$$\mathbf{s}_i^k = g(\hat{\mathbf{f}}_i^k, \mathbf{W}), \tag{3.11}$$

where $\mathbf{W}$ is the learnable weights of the shared MLP. Finally, the features are weighted and summed as follows:

$$\tilde{\mathbf{f}}_i^k = \sum_{k=1}^{K} (\hat{\mathbf{f}}_i^k \cdot \mathbf{s}_i^k). \tag{3.12}$$

In general, the LocSE aggregates the geometry information with the features of the local region, and attentive pooling learns to pool more informatively. Thus, the stacks of multiple LocSE and attentive pooling units increase the receptive field of each layer to learn local complex local structures.

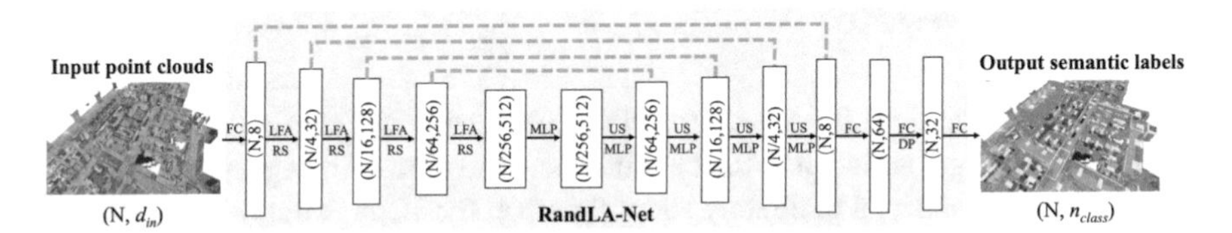

Fig. 3.11 RandLA-Net architecture overview. Reproduced with permission [11]. Copyright © 2020, IEEE

The details of the RandLA-Net architecture are shown in Fig. 3.11. The network follows the commonly used encoder–decoder architecture with skip connections. The input point cloud, which has a 10^5 scale with 3D coordinates and color, is fed into a shared MLP to extract per-point features. Then, four encoding layers are used to reduce the size of the point cloud to 10^2 while increasing the feature dimensions. Next, four decoding layers are employed using nearest neighbor interpolation to upsample the point cloud and concatenate with the features in the corresponding encoding layer. Finally, three fully connected (FC) layers and a dropout layer are used to obtain a semantic prediction.

RandLA-Net is 200-times faster than existing approaches and achieves the state-of-the-art semantic segmentation performance on the Semantic3D [10], SemanticKITTI [4], and S3DIS [3] datasets. For the S3DIS dataset, it obtains an overall accuracy of 88%, a mean accuracy of 82%, and mean IoU of 70% with six-fold cross validation.

3.3 Registration

The potential of the early works on deep networks for point cloud classification and semantic segmentation, such as PointNet and DGCNN, turned the attention of the research community to similar deep networks for other tasks, including point cloud registration. The iterative closest point (ICP) method for object registration (see Sect. 2.6.1) and its variants suffer from the problem of local minima. This means that they only perform well when the optimal alignment is close to the initial alignment; that is, the initial alignment needs to be good in order to obtain a tighter alignment. Some traditional methods search for a global solution, but they can be an order of magnitude slower. In this section, we discuss the works of PointNetLK, Deep Closest Point (DCP), PRNet, 3D Match, PPFNet, and Deep Global Registration. These methods utilize Deep learning to solve geometric registration problems, with the aim of overcoming some of the drawbacks of traditional methods. The networks are trained in an end-to-end manner with different forms of supervision.

3.3.1 PointNetLK

PointNet [18] successfully demonstrated the application of Deep learning for point cloud processing, particularly for classification and semantic segmentation tasks. However, it is nontrivial to directly adopt PointNet for 3D registration. PointNetLK [2] extends PointNet for point cloud registration by extracting the global point cloud feature vector of the two point clouds to be registered (source and template) using PointNet and combines it with the classical Lucas and Kanade (LK) algorithm [17] for iterative alignment. The authors considered PointNet from the perspective of a trainable imaging function, which motivated them to apply the well-established image registration LK algorithm to PointNet. However, due to the unordered nature of points and lack of well-defined neighborhoods in point clouds, it is not possible to compute local gradients, as demanded by the LK algorithm. Thus, a modified LK algorithm was proposed for point cloud processing.

The registration problem is set up as follows. The PointNet function ϕ encodes a K-dimensional global vector descriptor of the point cloud. This is the feature vector obtained after the global max pooling operation in the classification pipeline of PointNet. P_T and P_S are the template and source point clouds, respectively. The goal is to find a rigid body transformation $G \in \mathrm{SE}(3)$ (special Euclidean group in three dimensions) that best aligns P_S to P_T. G is represented as

$$G = \exp\left(\sum_i \xi_i T_i\right), \qquad \xi = (\xi_1, \xi_2, \ldots, \xi_6)^T, \tag{3.13}$$

where T_i are generators with twist parameters $\xi \in \mathbb{R}^6$. Taking this into consideration, PointNetLK tries to find the optimal G, such that the PointNet function of the template is equal to the PointNet function of the transformed source. That is, $\phi(P_T) = \phi(G \cdot P_S)$. The input and feature transformation modules (T-net) in PointNet are omitted in PointNetLK, and the fully connected layers after global pooling are not present, as these two elements are a part of the classification framework and not required for registration. Furthermore, to reduce the computational time of each iteration, an inverse compositional (IC) formulation is used, which reversed the role of template and source. In every iteration, incremental warps to the template are found and inverse of the transformation is applied to the source. This modifies the objective to

$$\phi(P_S) = \phi(G^{-1} \cdot P_T). \tag{3.14}$$

The right hand side of Eq. 3.14 is then linearized as

$$\phi(P_S) = \phi(P_T) + \frac{\partial}{\partial \xi}[\phi(G^{-1} \cdot P_T)]\, \xi, \tag{3.15}$$

where G^{-1} is given by $G = \exp\left(-\sum_i \xi_i T_i\right)$. Further, $J \in \mathbb{R}^{K\times 6}$ is the Jacobian matrix and $J = \frac{\partial}{\partial \xi}[\phi(G^{-1} \cdot P_T)]$. Unlike images, it is nontrivial to compute J for point clouds. Hence, in the modified LK algorithm, every column J_i is estimated using finite difference gradients as

$$J_i = \frac{\phi(\exp(-t_i T_i) \cdot P_T) - \phi(P_T)}{t_i}, \tag{3.16}$$

where t_i is a small perturbation of ξ. The solution for ξ in Eq. 3.15 is given by

$$\xi = J^{+}[\phi(P_S) - \phi(P_T)]. \tag{3.17}$$

Here, J^+ is the pseudo-inverse of matrix J. Using ξ, the one step update ΔG can be used to update the source point cloud as

$$P_S \leftarrow \Delta G \cdot P_S, \qquad \Delta G = \exp\left(\sum_i \xi_i T_i\right). \tag{3.18}$$

The final transformation G_{est} is a composition of all the transformation estimates found in every iteration and is given by

$$\Delta G_{est} = \Delta G_n \cdot \ldots \cdot \Delta G_1 \cdot \Delta G_0. \tag{3.19}$$

The loss function during training of PointNetLK uses the orthogonality property of the spatial transformation G and is given by

$$\|(G_{est})^{-1} \cdot G_{gt} - I_4\|_F, \tag{3.20}$$

where G_{gt} is the ground truth transformation matrix that is used for supervision, I_4 is an identity matrix, and $\|\cdot\|_F$ is the matrix Frobenius norm. The PointNetLK architecture is shown in Fig. 3.12. The one-time and looping computations are

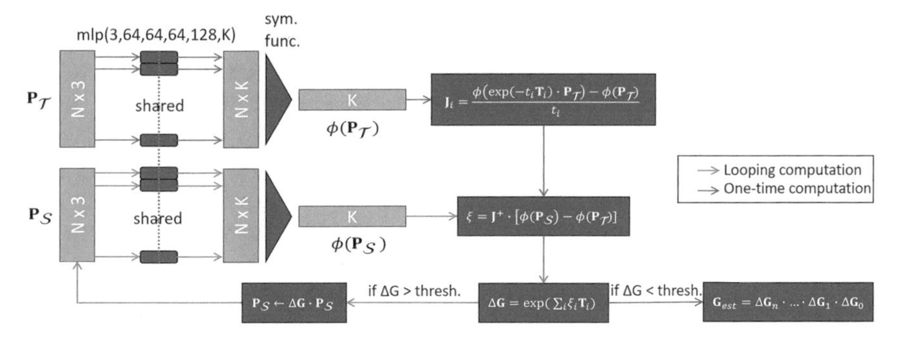

Fig. 3.12 PointNetLK architecture. Reproduced with permission [2]. Copyright © 2019, IEEE

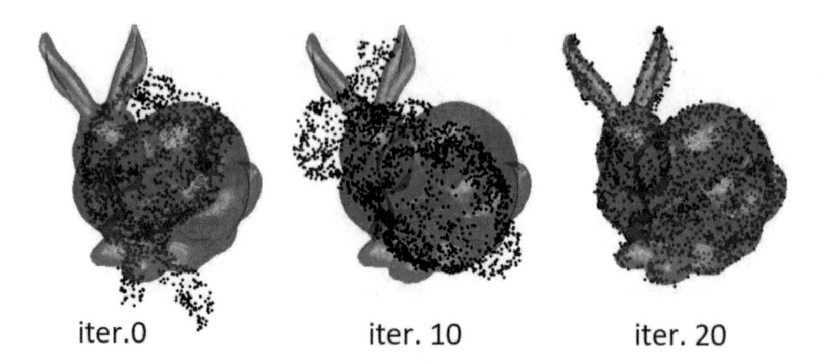

Fig. 3.13 Alignment using PointNetLK. Template and source after 0, 10, and 20 iterations. Reproduced with permission [2]. Copyright © 2019, IEEE

highlighted by the blue and orange lines, respectively. We can see that the Jacobian matrix is computed only once from the feature of the template point cloud. The twist parameters are updated in every iteration based on the incremental alignments of the source point cloud. A minimum threshold is set for ΔG which serves as the stopping criteria.

Extensive experiments on the ModelNet40 dataset have shown the effectiveness of PointNetLK for 3D object registration. In particular, PointNetLK performs well even with relatively large rotation angles, when the ICP method would nearly always fail. Furthermore, PointNetLK is robust to added Gaussian noise. For noise resilience, global average pooling works better than global max pooling, which is used in the original PointNet implementation. PointNetLK also generalizes well with regard to object categories that are excluded during training. An example of iterative registration of the Stanford bunny model is shown in Fig. 3.13.

3.3.2 *Deep Closest Point*

Similar to PointNet [18] and PointNet++ [19], DGCNN [25] was another pioneering work targeted at the point cloud classification and semantic segmentation tasks. The authors of DGCNN developed a registration method that closely mimics the ICP-like alignment pipeline; however, unlike the iterative path of ICP, they proposed a method that employed a one-pass end-to-end trained deep network capable of global registration. This method was termed Deep Closest Point (DCP) [23]. It is common for traditional methods (ICP and many more) to find point correspondences and use them to estimate the 3D transformation that best aligns two point clouds. DCP follows a similar correspondence-based approach. In contrast, as we saw in the Sect. 3.3.1, PointNetLK uses a global point cloud feature vector to find the transformation. The method used by DCP for registration can be summarized in four steps: (1) point feature extraction using DGCNN; (2) feature transformation using

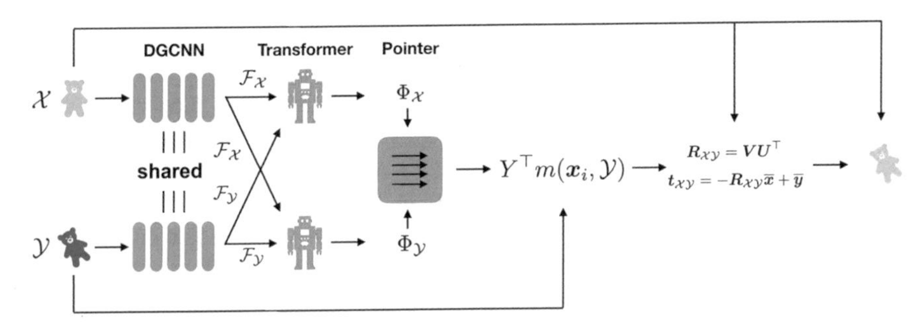

Fig. 3.14 DCP architecture. Reproduced with permission [23]. Copyright © 2019, IEEE

Transformer; (3) soft pointer generation using a pointer network; and (4) estimating the transformation using singular value decomposition (SVD).

The rigid alignment problem is formulated as follows. The two point clouds to be aligned are represented as $X = \{x_1, \ldots, x_i, \ldots, x_N\} \in \mathbb{R}^3$ and $Y = \{y_1, \ldots, y_i, \ldots, y_N\} \in \mathbb{R}^3$, with the assumption that both point clouds contain the same number of points N. A rigid transformation $[R_{XY}, t_{XY}]$ is applied to X in order to obtain Y, where $R_{XY} \in SO(3)$ is the 3D rotation matrix and $t_{XY} \in \mathbb{R}^3$ is the translation vector. The goal of the network is to minimize the error term $E(R_{XY}, t_{XY})$ which is given by

$$E(R_{XY}, t_{XY}) = \frac{1}{N}\sum_{i}^{N} \|R_{XY}x_i + t_{XY} - y_i\|^2, \tag{3.21}$$

where (x_i, y_i) are the ordered pairs of corresponding points.

Next, we consider the details of each step. The DCP architecture is shown in Fig. 3.14.

Point Feature Extraction In the point feature extraction step, the point clouds X, $Y \in \mathbb{R}^{(N\times 3)}$ are mapped to a higher dimensional feature embedding space using a trainable network to obtain the point features. For this, two learning modules, PointNet and DGCNN, can be used. PointNet extracts the features of each point independently, whereas DGCNN incorporates local geometry information in the feature learning process by constructing a graph of the neighboring points, applying a nonlinearity at the edge endpoints, and finally performing a vertex-wise local aggregation. The global max pooling operations in PointNet and DGCNN are avoided, as they are relevant only to the classification task. For registration, only pointwise features are used. At the output of this step, we obtain point features denoted by $F_X = \{x_1^L, x_2^L, \ldots, x_i^L, \ldots, x_N^L\}$ and $F_Y = \{y_1^L, y_2^L, \ldots, y_i^L, \ldots, y_N^L\}$, considering L layers in DGCNN or PointNet. Empirical results show that DGCNN performs better than PointNet for this step, mainly due to the incorporation of local structure information.

Feature Transformation The next step, feature transformation, captures self-attention and conditional attention. Transformer, which is based on the principle of attention, has shown massive success in natural language processing tasks like machine translation. So far, the features F_X and F_Y have been learned independently; that is, X does not influence F_Y and Y does not influence F_X. The attention network in the form of a transformer is included with the goal of capturing contextual information between two feature embeddings. The attention model learns a function $\phi : \mathbb{R}^{N\times P} \times \mathbb{R}^{N\times P} \rightarrow \mathbb{R}^{N\times P}$, where P is the dimension of the feature embedding from previous step. The new feature embeddings are then given by

$$\begin{aligned} \Phi_X &= F_X + \phi(F_X, F_Y), \\ \Phi_Y &= F_Y + \phi(F_Y, F_X). \end{aligned} \tag{3.22}$$

This operation modifies $F_X \rightarrow \Phi_X$ such that the point features of X have knowledge about the structure of Y. For function Φ, the transformer network [22] is incorporated using four head attention.

Pointer Generation After feature transformation, the pointer generation step is conducted. The goal of this step is to establish point correspondences. However, unlike a hard assignment, which is non-differentiable, DCP uses a probabilistic approach to generate soft pointers that allow the computation and propagation of gradients. Accordingly, each point $x_i \in X$ is assigned a probability vector over the points $y_i \in Y$ and is given by

$$m(x_i, Y) = softmax(\Phi_Y \Phi_{x_i}^T). \tag{3.23}$$

Here, Φ_{x_i} denotes the i-th row of Φ_X and $m(x_i, Y)$ is the soft pointer.

Transformation Estimation The final step is to predict the transformation, i.e., R_{XY} and t_{XY}. This is done using Singular Value Decomposition (SVD). First, the soft pointers are used to generate an estimate of the matching point in Y for every point in X. This is given by

$$\hat{y}_i = Y^T m(x_i, Y) \in \mathbb{R}^3, \tag{3.24}$$

where $Y \in \mathbb{R}^{N\times 3}$ is the matrix of input points. Then, $(x_i, \hat{y}_i)$ are used as point correspondences to find the transformation. First, the centroids of X and Y are found:

$$\bar{x} = \frac{1}{N}\sum_{i=1}^{N} x_i \quad \text{and} \quad \bar{y} = \frac{1}{N}\sum_{i=1}^{N} y_i. \tag{3.25}$$

Then, the covariance matrix is calculated as

$$H = \sum_{i=1}^{N} (x_i - \bar{x})(y_i - \bar{y})^T. \tag{3.26}$$

The matrix H is then decomposed as $H = USV^T$ using SVD, where $U \in SO(3)$ is the matrix of left singular vectors, S is the 3×3 diagonal matrix of singular values, and $V \in SO(3)$ is the matrix of right singular vectors. Thereafter, the optimal $[R_{XY}, t_{XY}]$ to minimize Eq. 3.21 is given by

$$R_{XY} = VU^T \quad \text{and} \quad t_{XY} = -R_{XY}\bar{x} + \bar{y}. \tag{3.27}$$

The error function in Eq. 3.21 assumes exact point correspondences are known. Since DCP uses soft pointers, only approximate locations in the Y point cloud are obtained instead. Hence, the error term can be modified to

$$E(R_{XY}, t_{XY}) = \frac{1}{N}\sum_{i}^{N} \|R_{XY}x_i + t_{XY} - y_{m(x_i)}\|^2. \tag{3.28}$$

The mapping function $m(\cdot)$ is learned with the objective

$$m(x_i, Y) = \arg\min_{j} \|R_{XY}x_i + t_{XY} - y_j\|. \tag{3.29}$$

The network is trained in an end-to-end manner using the ground truth rotation matrix and translation vector as supervision. The loss term is given by

$$Loss = \|R_{XY}^T R_{XY}^g - I\|^2 + \|t_{XY} - t_{XY}^g\|^2 + \lambda\|\theta\|^2, \tag{3.30}$$

where superscript g denotes the ground truth transformations and θ is a regularization term.

For training and evaluation, the ModelNet40 dataset is used. The experiments are mainly split into three parts—tests on unseen point clouds (test data), tests on unseen categories (object classes set aside during training), and registration of noisy point clouds. DCP outperforms PointNetLK as well as traditional methods like ICP. The inclusion of the Transformer module further reduces the error.

3.3.3 PRNet

PointNetLK and DCP were instrumental with regard to their early impact on the 3D point cloud research community. However, the main assumption that they made was that complete point clouds would be available during alignment. In practice, only partial views are visible and only a subset of points of the source and target point clouds overlap. This limits the application of PointNetLK and DCP when using real-world scans. In their follow-up work, the authors of DCP proposed a network that can handle partial point cloud registration. This partial registration network is termed PRNet [24]. PRNet facilitates sequential alignment that enables

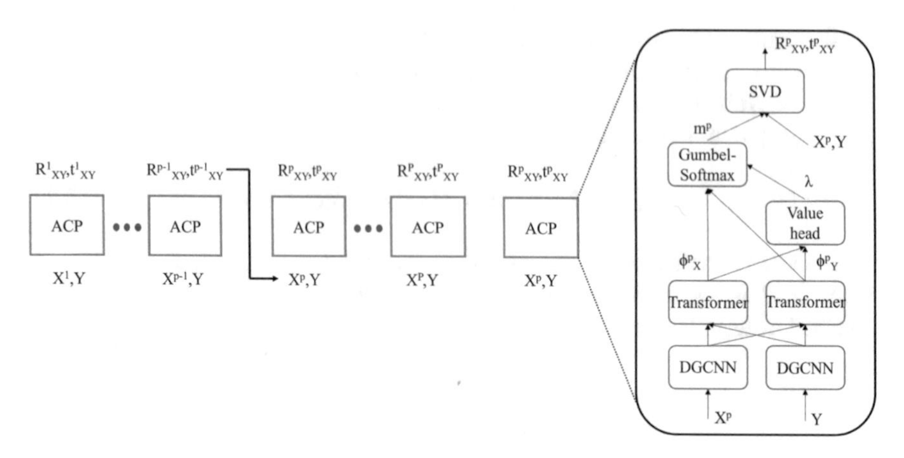

Fig. 3.15 PRNet architecture. ACP stands for Actor–Critic Close Point module

coarse-to-fine refinement of an initial estimate. The key element of PRNet is a sub-module that identifies matching keypoints in two point clouds using co-contextual information. Furthermore, the learned representations can be transferred to point cloud classification tasks by training a classifier. Unlike DCP, which registers two point clouds in a one-pass manner, PRNet is an iterative method, similar to ICP.

The model architecture of PRNet is illustrated in Fig. 3.15. The problem formulation is very similar to that of DCP. $X = \{x_1, x_2, \ldots, x_i, \ldots, x_N\} \in \mathbb{R}^3$ and $Y = \{y_1, y_2, \ldots, y_i, \ldots, y_N\} \in \mathbb{R}^3$ are the two point clouds to be registered, and the goal is to find the rotation matrix $R_{XY} \in SO(3)$ and translation vector $t_{XY} \in \mathbb{R}^3$ that optimally aligns point cloud X to Y. The objective function is given by

$$E(R_{XY}, t_{XY}, m) = \frac{1}{N} \sum_{i}^{N} \| R_{XY} x_i + t_{XY} - y_{m(x_i)} \|^2, \tag{3.31}$$

where x_i are the points in X and $y_{m(x_i)}$ are the predicted points of Y using the soft pointer function $m(\cdot)$.

The registration process in PRNet can be summarized as follows. First, the input point clouds X and Y are taken and their keypoints are detected. Next, a mapping from the keypoints of X to the keypoints of Y is predicted. Further, using the keypoints and mapping, a rigid transformation $[R_{XY}, t_{XY}]$ is predicted that best aligns X to Y. Point cloud X is then transformed using the transformation. The process is repeated by using the transformed X, given by $(R_{XY} X + t_{XY})$ and Y.

$X^p = \{x_1^p, x_2^p, \ldots, x_i^p, \ldots, x_N^p\}$ is the transformed X point cloud at the p-th iteration in the alignment process. Similarly, $[R_{XY}^p, t_{XY}^p]$ is the p-th rigid motion predicted. The first step of keypoint detection is as follows. The authors associate the importance of point with the L^2 norm of point features. Points corresponding to features with a higher L^2 norm are selected as the keypoints. The k keypoints X_k^p and Y_k^p for point clouds X^p and Y^p are given by

$$X_k^p = X^p(\text{topk}(\|\Phi_{x_1}^p\|_2, \ldots, \|\Phi_{x_i}^p\|_2, \ldots \|\Phi_{x_N}^p\|_2))$$
$$Y_k^p = Y^p(\text{topk}(\|\Phi_{y_1}^p\|_2, \ldots, \|\Phi_{y_j}^p\|_2, \ldots \|\Phi_{y_N}^p\|_2)), \tag{3.32}$$

where $topk(\cdot)$ gives the points with the largest feature L^2 norm. Φ is the feature embedding obtained using DGCNN and Transformer. This embedding is the same as that used in DCP. The authors argue that the selected keypoints are common points between the two partial point clouds; hence, non-overlapping points are avoided during the transformation estimation step. The network learns to detect reasonable keypoints without any specific supervision.

PRNet adopts a slightly different approach to DCP for generating the set of point correspondences. The mapping function of DCP, as presented in Eq. 3.23, is differentiable, but smooth enough to make the mapping quite blurred. In contrast, PRNet uses a rather sharp and differentiable mapping function in the form of Gumbel-Softmax. The mapping at the p-th step is given by

$$m^p(x_i, Y) = one\ hot\left[\arg\max_j softmax\left(\phi_Y^p \phi_{x_i}^{p\,T} + g_{ij}\right)\right], \tag{3.33}$$

where g_{ij} are independent samples from the Gumbel(0,1) distribution. Next, we discuss the Actor–Critic Close Point (ACP) module. The main purpose of ACP is to adapt the sharpness of the mapping function in Eq. 3.33, with the aim of adjusting the alignment goals for different iterations (different values of p). In this way, a coarser alignment can be used in the initial steps, with a sharper mapping function that aligns individual corresponding points in the final steps. A parameter λ is added to Eq. 3.33 to produce a generalized matching matrix:

$$m^p(x_i, Y) = one\ hot\left[\arg\max_j softmax\left(\frac{\phi_Y^p \phi_{x_i}^{p\,T} + g_{ij}}{\lambda}\right)\right]. \tag{3.34}$$

With larger values of λ, the matrix is smoothed, whereas with smaller values of λ, it is closer to a binary matrix. A separate network Θ is used that considers the global point cloud feature vectors ψ_X^p and ψ_X^p to predict λ. The global point cloud feature is obtained by aggregating point features using average pooling as $\psi_X^p = avg_i(\phi_{x_i}^p)$ and $\psi_Y^p = avg_i(\phi_{y_i}^p)$. Then, $\lambda = \Theta(\psi_X^p, \psi_X^p)$. The actor–critic terminology is borrowed from reinforcement learning. The actor head outputs the rigid motion based on the λ value predicted by the critic head.

The loss function used is a combination of three losses—rigid motion loss L_p^m, cycle consistency loss L_p^c, and global feature alignment loss L_p^g. There is an additional parameter $\gamma \leq 1$ that promotes alignment in the initial steps. The loss is given by

$$L = \sum_{p=1}^{P} \gamma^{p-1} L_p, \text{ where } L_p = L_p^m + \alpha L_p^c + \alpha L_p^g. \tag{3.35}$$

Experiments on both synthetic ModelNet40 and real-world bunny dataset highlight the effectiveness of PRNet for the registration of partial 3D objects.

3.3.4 3D Match

3DMatch [31] employs Deep learning to learn a mapping function ψ that encodes a feature descriptor for a local 3D patch around a point. A smaller l_2 distance between features of two patches is desired for corresponding points (or patches). To generate pairs of correspondence for training the network, correspondence labels from RGB-D reconstruction datasets such as 7-Scenes [21] and SUN3D [27] are used. We encourage the reader to refer to the paper [31] for a better understanding of the details regarding ground truth correspondence generation.

The mapping function ψ in 3DMatch is a 3D ConvNet that outputs a 512-dimensional patch descriptor. The network weights are learned with the objective of minimizing the l_2 distance between the feature descriptors of corresponding patches, as well as maximizing the l_2 distance between the features of non-matching patches. The network architecture is illustrated in Fig. 3.16. It is a Siamese style 3D CNN comprising eight convolution layers and a pooling layer. It takes two local patches in the form of $30 \times 30 \times 30$ truncated distance function (TDF) voxel grids, as explained in the following paragraph, and predicts whether they correspond. Equal numbers of true matches and non-matches are fed to the network during training.

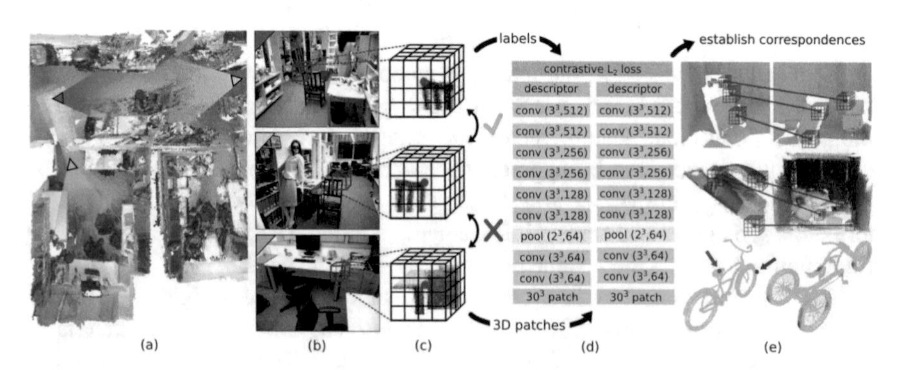

Fig. 3.16 3DMatch architecture. Reproduced with permission [31]. Copyright © 2017, IEEE. (**a**) RGB-D reconstruction. (**b**) Frames. (**c**) Local patches. (**d**) Siamese network. (**e**) Applications

Next, we go over the process of 3D data representation, which provides the input for the 3D CNN. For an input point, the 3D region in the local neighborhood is converted to a $30 \times 30 \times 30$ voxel grid of TDF values. The TDF value of every voxel represents the distance between the voxel center and the nearest 3D surface. Later, the TDF values are truncated and normalized. They lie between 0 and 1, where 1 is at the surface and 0 is far from the surface. These voxels are then aligned with respect to the camera view. The voxel representation, being an ordered grid, enables the use of 3D CNNs to process the patches.

The 3DMatch network has been evaluated for keypoint matching tasks as well as geometric registration tasks. For registration, the correspondences found using 3DMatch were combined with RANSAC for robust alignment of the 3D point clouds. For the matching task, 3DMatch outperforms handcrafted features such as Spin-Images [13] and FPFH [20]. Further experiments also show that 3DMatch can be integrated into 3D reconstruction frameworks. That is, the keypoint matches can be considered during the bundle adjustment step to refine the reconstructed 3D model.

3DMatch differs from the other methods discussed in this book, in that it converts unordered points in a local region (local patch) to a regular 3D representation in the form of a voxel grid. In contrast, the other discussed methods process the 3D points directly, without converting them to voxels or any other ordered representation. There are a wide range of voxel-based (or mesh-based) methods, which are beyond the scope of this book.

3.3.5 PPFNet

PPFNet [7] employs a globally informed 3D local feature descriptor to find correspondences between two point clouds. It takes a local point cloud patch and encodes its point-pair-feature along with raw point coordinates and point normal information. Further, PointNet is used to learn permutation invariant point features. The final feature is found by concatenating the local point features and global point cloud features, followed by a further MLP layer. The mechanism of PPFNet is discussed in detail below.

PPFNet considers two point clouds, $X \in \mathbb{R}^3$ and $Y \in \mathbb{R}^3$, with x_i and y_i being their i^{th} points. Assuming a rigid transformation, X and Y are related by a permutation matrix $P \in \mathbb{P}^n$ and a rigid transformation $T = R \in SO(3), t \in \mathbb{R}^3$. The error for registration is given by

$$d(X, Y|R, t, P) = \frac{1}{N} \sum_{i=1}^{n} \|x_i - y_{i(P)} - t\|^2. \tag{3.36}$$

In matrix notation,

$$d(X, Y|R, t, P) = \frac{1}{N}\|X - PYT^T\|^2. \tag{3.37}$$

PPFNet attempts to learn an effective mapping function $f(\cdot)$ that achieves $d_f(X, Y|T, P) \approx 0$ under any transformation T and permutation P, and,

$$d_f(X, Y|R, t, P) = \frac{1}{N}\|f(X) - f(PYT^T)\|^2. \tag{3.38}$$

The function f is designed to be invariant to point permutations P and tolerant to transformations T.

Next, we go over the point pair features (PPF) method. Given two 3D points x_1 and x_2, their PPF, ψ_{12}, defines the surface characteristics as follows:

$$\psi_{12} = (\|d\|_2, \angle(n_1, d), \angle(n_2, d), \angle(n_1, n_2)), \tag{3.39}$$

where $\|d\|_2$ is the l_2 distance between the two points; and, n_1 and n_2 are the surface normals of x_1 and x_2, respectively. $\angle$ is the angle between two directions that lies in $[0, \pi)$ and computed as

$$\angle(v_1, v_2) = \text{atan2}(\|v_1 \times v_2\|, v_1 \cdot v_2). \tag{3.40}$$

PPF is invariant under 3D rotation, translation, and reflection.

We next study the local geometry encoding process of PPFNet, which forms the basis of the network. For a reference point $x_r \in X$, a set of points $\{m_i\} \in \Omega \subset X$ in the local neighborhood is found. The local reference frame is found around x_r with respect to the neighboring points in Ω. Then for every point m_i in the set Ω, its PPF ψ_{ri} is found by pairing it with the reference point x_r. The PPFs are concatenated with the point coordinates and point normals to obtain the local geometry encoding of the reference point. This is given by

$$F_r = \{x_r, n_r, x_i, \cdots, n_i, \cdots, \psi_{ri}, \cdots\}. \tag{3.41}$$

The local geometry encoding is shown in Fig. 3.17.

After encoding the local patch information, PointNet [18] is used to extract features from the local patches. N local patches are uniformly sampled from the point cloud and presented to the network. Max pooling is used to aggregate the features of all local patches into a global descriptor for the entire point cloud. The local patch features and global pooled features are then concatenated to further learn the final patch features. The feature construction process is illustrated in Fig. 3.18.

The network is trained using an N-tuple loss that considers N pairs of patches from the two point clouds to be matched. By selecting some pairs of true correspondences and some pairs of non-corresponding patches, the network learns distinctive

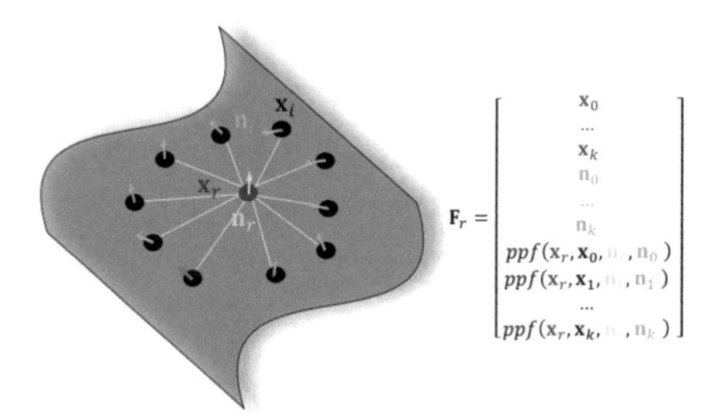

Fig. 3.17 Local geometry encoding in PPFNet. Reproduced with permission [7]. Copyright © 2018, IEEE

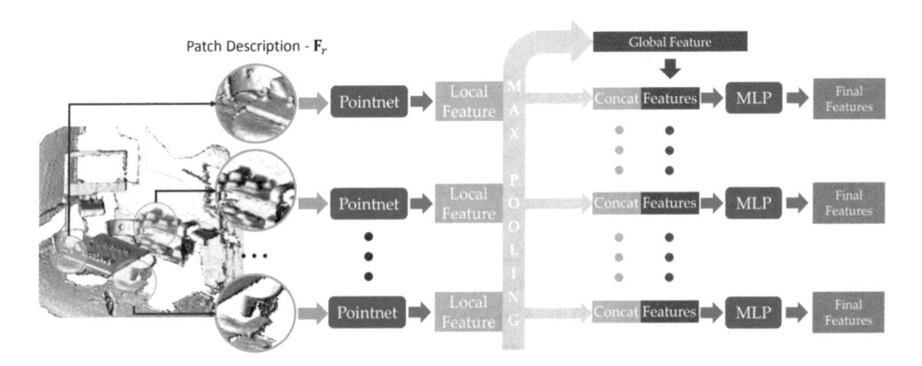

Fig. 3.18 PPFNet patch feature construction. Reproduced with permission [7]. Copyright © 2018, IEEE

features that preserve proximity in feature space for matching patches. These set of true and false correspondences are selected using ground truth pose information. The detailed architecture is shown in Fig. 3.19.

The performance of PPFNet for matching and geometric registration tasks has been evaluated on the 3DMatch dataset. PPFNet outperforms 3DMatch and several other handcrafted 3D descriptors.

3.3.6 Deep Global Registration

A wide range of point cloud registration methods employ a similar general framework, which can be summarized as follows. First, local geometric features are extracted for all points. These features can be handcrafted (as in traditional methods) or learned using end-to-end Deep learning. Next, point correspondences are found using a nearest neighbors rule in the feature space. Typically, different criteria such as ratio or reciprocity tests are used to filter out the correspondences. Finally, the

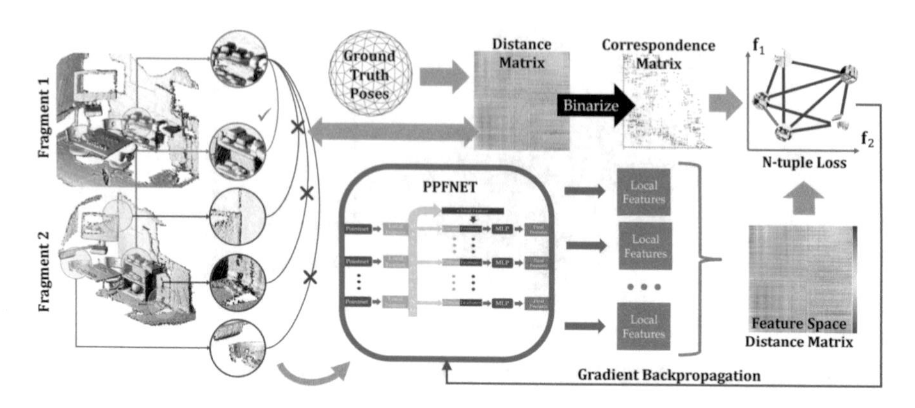

Fig. 3.19 PPFNet architecture. Reproduced with permission [7]. Copyright © 2018, IEEE

transformation is estimated from the set of corresponding points. Although this is the general pipeline followed, some methods like PointNetLK [2] deviate from this general approach.

Deep Global Registration [5] goes further and proposes a learning-based network to predict the confidence of a pair of point correspondences. These confidence scores are then used in a differentiable weighted Procrustes algorithm to predict the 3D transformation. Finally, a gradient-based optimizer is proposed for pose refinement. The major contribution of this method lies in the steps after finding an initial set of point correspondences. Any of the methods discussed so far can be used to find pointwise local features. However, a fully convolutional geometric features (FCGF) [6] method is typically used experimentally. Deep Global Registration has advantages both in terms of registration accuracy and computation efficiency. For instance, the weighted Procrustes method reduces the optimization complexity from quadratic to linear time.

The process of correspondence confidence prediction is as follows. $F_x = \{f_{x_1}, \ldots, f_{x_{N_x}}\}$ and $F_y = \{f_{y_1}, \ldots, f_{y_{N_y}}\}$ are considered to be the pointwise features of the two point clouds to be registered. Here, N_x and N_y represent the number of points in point cloud X and Y, respectively. An initial set of correspondences i, j is found using nearest neighbor rule $M = \{i, \arg\min_j \| f_{x_i} - f_{y_j} \|)i \in [1, \ldots, N_x]\}$. In a typical setting, this initial set is examined using some handcrafted approaches such as ratio test to select a subset of inlier correspondences. The authors use a convolutional network for this task, thereby eliminating the need for a heuristic outlier rejection method. The role of the convolutional network is to analyze the geometric structure of the correspondence set.

The input to the convolutional network is a set of vectors in $\mathbb{R}^6$ that are formed by concatenation of the 3D coordinates of the corresponding points $x_i \leftrightarrow y_j$ and is given by $[x_i^T, y_j^T]^T \in \mathbb{R}^6$. The authors hypothesize that the inlier correspondences lie on a lower-dimensional surface in 6D space governed by the 3D geometry of the input. P is used to represent the set of inlier correspondences and is given by

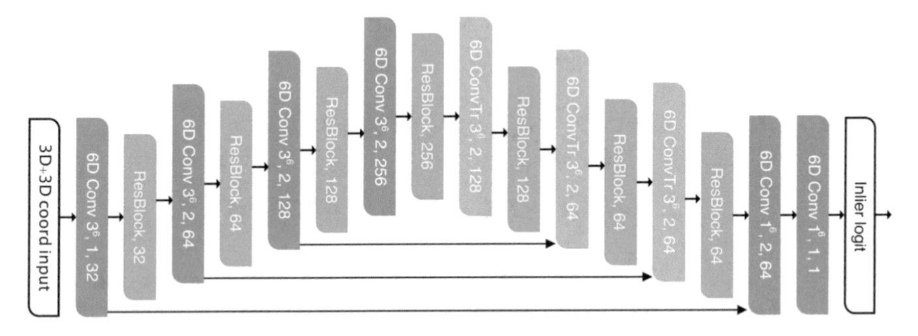

Fig. 3.20 Convolutional network for inlier likelihood prediction. Reproduced with permission [5]. Copyright © 2020, IEEE

$P = \{(i, j) | \, \|T^*(x_i) - y_j\| < \tau, (i, j) \in M\}$. This is a subset of correspondences (i, j) from the initial set M that align under a threshold τ and with ground truth transformation T^*. $N = P^C \cap M$ is used to denote the set of outlier correspondences. The 6D ConvNet outputs a likelihood that the correspondence is an inlier. The convolutional network for inlier likelihood prediction is illustrated in Fig. 3.20. It consists of a U-Net-like structure with residual blocks.

Binary cross-entropy loss between the likelihood prediction of inliers and ground truth correspondences is used during training. It is given by

$$L_{bce}(M, T^*) = \frac{1}{|M|} \sum_{(i,j)\in P} \log p_{(i,j)} + \sum_{(i,j)\in N} \log (1 - p_{(i,j)}), \tag{3.42}$$

where $|M|$ is the number of initial correspondences.

In the following step, the likelihood predictions are used to estimate the transformation. This is achieved using the weighted Procrustes method. The original Procrustes method, such as that used in ICP and DCP, gives equal weight to every correspondence. The minimization criterion used there is the mean-squared error between corresponding points, which is given by

$$\frac{1}{N} \sum_{(i,j)\in M} \|x_i - y_j\|. \tag{3.43}$$

The weighted mean-squared error function that assigns a different weight to each correspondence is given by

$$\sum_{(i,j)\in M} w_{(i,j)} \|x_i - y_j\|. \tag{3.44}$$

Therefore, the weighted Procrustes method in Deep Global Registration minimizes the squared error as follows:

$$\begin{aligned} e^2(R, t; w, X, Y) &= \sum_{(i,j)\in M} w_{\bar{(i,j)}}(y_j - (Rx_i + t))^2 \\ &= tr((Y - RX - \mathbf{t1^T})\mathbf{W}(\mathbf{Y} - \mathbf{RX} - \mathbf{t1^T})^\mathbf{T}), \end{aligned} \tag{3.45}$$

where $\mathbf{1} = [1, \ldots, 1]^T$, $w = [w_1, \ldots, w_{|M|}]$ is the weight vector and $\bar{w} = [\bar{w_1}, \ldots, w_{\bar{|M|}}]$ is the normalized weight vector obtained after applying a nonlinear transformation ϕ to w. W is the diagonal weight matrix whose diagonal elements are $\bar{w}$. The optimal transformation is then given by

$$\begin{aligned} \hat{R} &= USV^T \\ \hat{t} &= (Y - RX)W\mathbf{1}. \end{aligned} \tag{3.46}$$

Here, USV^T is the SVD of Σ_{xy}, $\Sigma_{xy} = YKWKX^T$, $K = I - \sqrt{\bar{w}}\sqrt{\bar{w}}^T$, and $S = diag(1, \ldots, 1, \det(U)\det(V))$.

Finally, a registration fine-tuning module is employed that uses a gradient-based method to minimize the loss function and improve registration accuracy. The details of this module are omitted here.

References

1. Abid, A., Balin, M.F., Zou, J.: Concrete autoencoders for differentiable feature selection and reconstruction (2019). arXiv preprint arXiv:1901.09346
2. Aoki, Y., Goforth, H., Srivatsan, R.A., Lucey, S.: Pointnetlk: Robust & efficient point cloud registration using pointnet. In: Proceedings of the IEEE/CVF Conference on Computer Vision and Pattern Recognition, pp. 7163–7172 (2019)
3. Armeni, I., Sener, O., Zamir, A.R., Jiang, H., Brilakis, I., Fischer, M., Savarese, S.: 3d semantic parsing of large-scale indoor spaces. In: Proceedings of the IEEE Conference on Computer Vision and Pattern Recognition, pp. 1534–1543 (2016)
4. Behley, J., Garbade, M., Milioto, A., Quenzel, J., Behnke, S., Stachniss, C., Gall, J.: Semantickitti: A dataset for semantic scene understanding of LiDAR sequences. In: Proceedings of the IEEE/CVF International Conference on Computer Vision, pp. 9297–9307 (2019)
5. Choy, C., Dong, W., Koltun, V.: Deep global registration. In: Proceedings of the IEEE/CVF Conference on Computer Vision and Pattern Recognition, pp. 2514–2523 (2020)
6. Choy, C., Park, J., Koltun, V.: Fully convolutional geometric features. In: Proceedings of the IEEE/CVF International Conference on Computer Vision, pp. 8958–8966 (2019)
7. Deng, H., Birdal, T., Ilic, S.: PPFNet: Global context aware local features for robust 3d point matching. In: Proceedings of the IEEE Conference on Computer Vision and Pattern Recognition, pp. 195–205 (2018)
8. Dovrat, O., Lang, I., Avidan, S.: Learning to sample. In: Proceedings of the IEEE/CVF Conference on Computer Vision and Pattern Recognition, pp. 2760–2769 (2019)

9. Groh, F., Wieschollek, P., Lensch, H.: Flex-convolution (million-scale point-cloud learning beyond grid-worlds) (2018). arXiv preprint arXiv:1803.07289
10. Hackel, T., Savinov, N., Ladicky, L., Wegner, J.D., Schindler, K., Pollefeys, M.: Semantic3d. net: a new large-scale point cloud classification benchmark (2017). arXiv preprint arXiv:1704.03847
11. Hu, Q., Yang, B., Xie, L., Rosa, S., Guo, Y., Wang, Z., Trigoni, N., Markham, A.: Randla-net: efficient semantic segmentation of large-scale point clouds. In: Proceedings of the IEEE/CVF Conference on Computer Vision and Pattern Recognition, pp. 11108–11117 (2020)
12. Jiang, M., Wu, Y., Zhao, T., Zhao, Z., Lu, C.: PointSIFT: A sift-like network module for 3d point cloud semantic segmentation (2018). arXiv preprint arXiv:1807.00652
13. Johnson, A.E.: Spin-images: a representation for 3-d surface matching (1997)
14. Krizhevsky, A., Sutskever, I., Hinton, G.E.: ImageNet classification with deep convolutional neural networks. Adv. Neural Inform. Process. Syst. **25**, 1097–1105 (2012)
15. Li, Y., Bu, R., Sun, M., Wu, W., Di, X., Chen, B.: PointCNN: convolution on x-transformed points. Adv. Neural Inform. Process. Syst. **31**, 820–830 (2018)
16. Lowe, D.G.: Distinctive image features from scale-invariant keypoints. Int. J. Comput. Vis. **60**(2), 91–110 (2004)
17. Lucas, B.D., Kanade, T., et al.: An iterative image registration technique with an application to stereo vision. Vancouver, British Columbia (1981)
18. Qi, C.R., Su, H., Mo, K., Guibas, L.J.: Pointnet: Deep learning on point sets for 3d classification and segmentation. In: Proceedings of the IEEE Conference on Computer Vision and Pattern Recognition, pp. 652–660 (2017)
19. Qi, C.R., Yi, L., Su, H., Guibas, L.J.: Pointnet++: Deep hierarchical feature learning on point sets in a metric space (2017). arXiv preprint arXiv:1706.02413
20. Rusu, R.B., Blodow, N., Beetz, M.: Fast point feature histograms (FPFH) for 3d registration. In: 2009 IEEE International Conference on Robotics and Automation, pp. 3212–3217. IEEE, Piscataway (2009)
21. Shotton, J., Glocker, B., Zach, C., Izadi, S., Criminisi, A., Fitzgibbon, A.: Scene coordinate regression forests for camera relocalization in RGB-D images. In: Proceedings of the IEEE Conference on Computer Vision and Pattern Recognition, pp. 2930–2937 (2013)
22. Vaswani, A., Shazeer, N., Parmar, N., Uszkoreit, J., Jones, L., Gomez, A.N., Kaiser, L., Polosukhin, I.: Attention is all you need (2017). arXiv preprint arXiv:1706.03762
23. Wang, Y., Solomon, J.M.: Deep closest point: Learning representations for point cloud registration. In: Proceedings of the IEEE/CVF International Conference on Computer Vision, pp. 3523–3532 (2019)
24. Wang, Y., Solomon, J.M.: PRNet: Self-supervised learning for partial-to-partial registration (2019). arXiv preprint arXiv:1910.12240
25. Wang, Y., Sun, Y., Liu, Z., Sarma, S.E., Bronstein, M.M., Solomon, J.M.: Dynamic graph CNN for learning on point clouds. ACM Trans. Graph. **38**(5), 1–12 (2019)
26. Wu, Z., Song, S., Khosla, A., Yu, F., Zhang, L., Tang, X., Xiao, J.: 3d ShapeNets: A deep representation for volumetric shapes. In: Proceedings of the IEEE Conference on Computer Vision and Pattern Recognition, pp. 1912–1920 (2015)
27. Xiao, J., Owens, A., Torralba, A.: Sun3d: A Database of Big Spaces Reconstructed Using SFM and Object Labels. In: Proceedings of the IEEE International Conference on Computer Vision, pp. 1625–1632 (2013)
28. Xu, K., Ba, J., Kiros, R., Cho, K., Courville, A., Salakhudinov, R., Zemel, R., Bengio, Y.: Show, attend and tell: Neural image caption generation with visual attention. In: International Conference on Machine Learning, pp. 2048–2057. PMLR (2015)
29. Yang, J., Zhang, Q., Ni, B., Li, L., Liu, J., Zhou, M., Tian, Q.: Modeling point clouds with self-attention and gumbel subset sampling. In: Proceedings of the IEEE/CVF Conference on Computer Vision and Pattern Recognition, pp. 3323–3332 (2019)

30. Yi, L., Kim, V.G., Ceylan, D., Shen, I.C., Yan, M., Su, H., Lu, C., Huang, Q., Sheffer, A., Guibas, L.: A scalable active framework for region annotation in 3d shape collections. ACM Trans. Graph. **35**(6), 1–12 (2016)
31. Zeng, A., Song, S., Nießner, M., Fisher, M., Xiao, J., Funkhouser, T.: 3dmatch: Learning local geometric descriptors from RGB-D reconstructions. In: Proceedings of the IEEE Conference on Computer Vision and Pattern Recognition, pp. 1802–1811 (2017)
32. Zhao, H., Jiang, L., Jia, J., Torr, P., Koltun, V.: Point transformer (2020). arXiv preprint arXiv:2012.09164

Chapter 4
Explainable Machine Learning Methods for Point Cloud Analysis

Abstract Explainable machine learning methods for point cloud analysis aim to decrease the model and computation complexity of current methods while improving their interpretation. These methods are an extension of successive subspace learning (SSL) from 2D images to 3D point clouds. SSL offers a lightweight unsupervised feature learning method based on the inherent statistical properties of data units. The model is significantly smaller than deep neural networks (DNNs) and more computationally efficient. However, it is nontrivial to generalize it to tackle the point cloud analysis problems, because points in a point cloud are irregular and unordered by nature, which is quite different from regular 2D images. In this chapter, we first discuss some early works on SSL processing of 2D images, then illustrate our explainable machine learning methods for point cloud classification, part segmentation, and registration in detail. Finally, we introduce some other applications of SSL.

4.1 Successive Subspace Learning on 2D Images

Deep learning is a black-box tool with extremely high training costs. To unveil its mysteries and reduce its complexity, a sequence of research work on SSL has been conducted by Professor Kuo and his students at the University of Southern California in the last 5 years, including [7–9, 18–20, 23, 27]. This series of works lays the foundation for explainable machine learning methods for point cloud analysis. SSL offers a lightweight unsupervised feature learning method based on the inherent statistical properties of data units. The model size is significantly smaller than that of DNNs and it is more computationally efficient. SSL has been applied to different data types such as images and point clouds, and it has proved to be effective in different applications such as image classification, facial recognition, point cloud registration, and so on.

In this section, we will introduce some early works on using SSL to analyze 2D images to explain the core design principle. Back in 2016, Kuo [18] noted that there was an issue with sign confusion arising from the cascade of hidden layers in convolutional neural networks (CNNs) and argued the need for nonlinear activation

S. Liu et al., *3D Point Cloud Analysis*,
https://doi.org/10.1007/978-3-030-89180-0_4

to eliminate this problem. Kuo [19] later noted that all filters in one convolutional layer form a subspace, which means each convolutional layer corresponds to a subspace approximation of the input. However, the analysis of subspace approximation is still complicated due to the existence of nonlinear activation. Therefore, it is desirable to solve the sign confusion problem by other means. The Saak (Subspace Approximation with Augmented Kernels) [9, 19] and the Saab (Subspace Approximation with Adjusted Bias) [20] transforms were proposed to avoid sign confusion while simultaneously preserving the subspace spanned by the filters fully.

4.1.1 Data-Driven Saak Transform

As its name suggests, the Saak transform [19] consists of two components: subspace approximation and kernel augmentation. To build the optimal linear subspace approximation, the second-order statistics of the input vectors are analyzed and the orthonormal eigenvectors of the covariance matrix are selected as transform kernels. This is the Karhunen–Loéve transform (KLT). Since the complexity of KLT increases dramatically when the input dimension is large, the Saak transform first decomposes high-dimensional vectors into multiple low-dimensional sub-vectors and repeats the process recursively. However, there is a sign confusion problem if two or more transforms are cascaded directly. A rectified linear unit (ReLU) is inserted between to solve the problem, which introduces rectification loss. To eliminate this loss, kernel augmentation is proposed by augmenting each transform kernel with its negative vector. Both the original and augmented kernels are used. With ReLU, one half of a transformed pair will pass through and the other half of the pair will be blocked. The integration of kernel augmentation and ReLU is equivalent to the sign-to-position (S/P) format conversion. Multiple Saak transforms are then cascaded to transform images of a larger size. The multistage Saak transforms offer a family of joint spatial–spectral representations between the full spatial domain representation and the full spectral domain representations.

An overview of the multistage Saak transform is presented in Fig. 4.1. Images are decomposed into four quadrants recursively to form a quadtree structure with its root being the full image and its leaves being small patches of size 2×2 pixels. The first-stage Saak transform is applied at the leaf node. Then, multistage Saak transforms are applied from all leaf nodes to their parents, stage by stage, until the root node is reached. Specifically, KLT is conducted on non-overlapping local cuboids of size $2 \times 2 \times K_0$ at the first stage, where $K_0 = 1$ for a monochrome image and $K_0 = 3$ for a color image. The horizontal and vertical spatial dimensions of the input image are reduced by one half. Thereafter, the KLT coefficients are augmented so that the spectral dimension is doubled to give $2^3 K_0$. In the next stage, KLT is conducted on non-overlapping local cuboids of size $2 \times 2 \times 2^3 K_0$, which yields an output of spectral dimension $K_2 = 2^3 K_1$ by kernel augmentation. The whole process is stopped when the kernel size reaches $1 \times 1 \times K_f$. If the image is of size $2^P \times 2^P$, we have $K_f = 2^{3P}$.

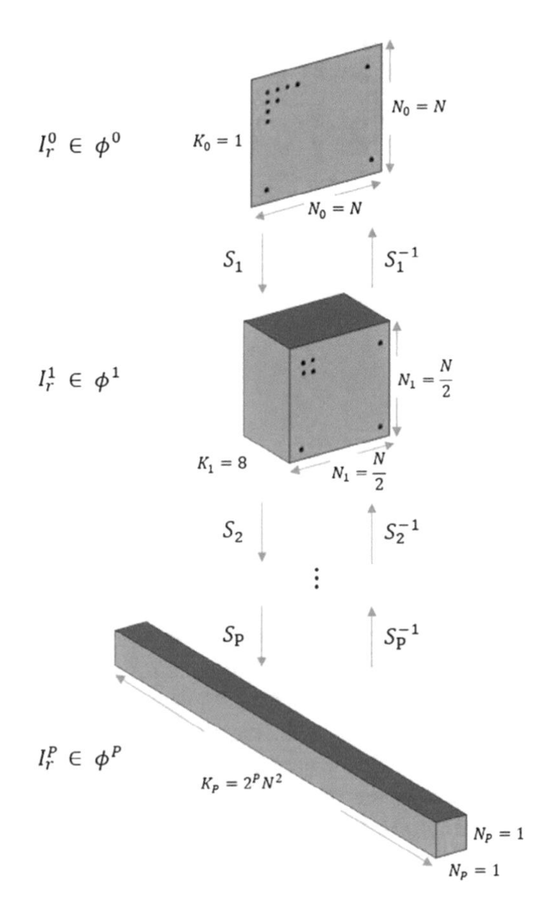

Fig. 4.1 Overview of the multistage Saak transform. Reproduced with permission [19]. Copyright © 2017, Elsevier Inc.

The Saak transform allows both forward and inverse transforms. This means that it can be used for image analysis as well as synthesis (or generation). The inverse Saak transform is conducted by performing a position-to-sign (P/S) format conversion before the inverse KLT. In general, the Saak transform converts the spatial variation to the spectral variation, while the inverse Saak transform converts the spectral variation to the spatial variation.

4.1.2 Handwritten Digit Recognition by Saak Transform

The Saak transform [9] was further improved to be more efficient, scalable, and robust to the handwritten digits recognition problem in this work. First, the principal component analysis (PCA) is adopted to select a smaller set of transform kernels. This results in a lossy Saak transform, which provides greater control over the size of the Saak coefficients and higher efficiency. Furthermore, because the feature

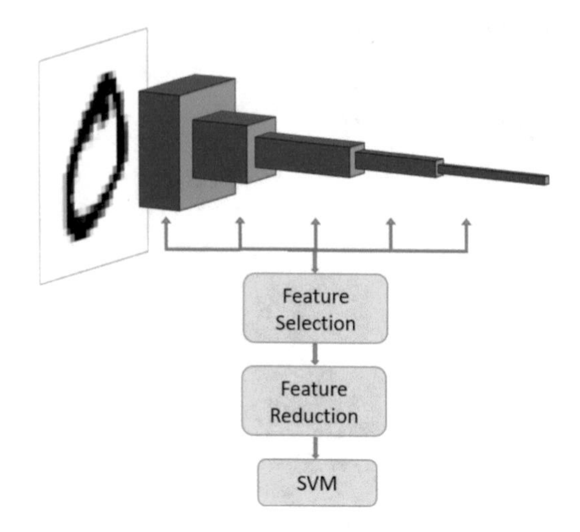

Fig. 4.2 Illustration of the Saak transform approach for pattern recognition. Reproduced with permission [9]. Copyright © 2018, IEEE

extraction process in the Saak transform approach is unsupervised and not sensitive to the class number, this method has good scalability. Finally, because only the principal components are kept, the influence of small perturbations is alleviated, making this improved method more robust.

An overview of the Saak transform approach for pattern recognition is shown in Fig. 4.2. First, a multistage Saak transform is used to extract a family of joint spatial–spectral representations of input images. Then, the Saak coefficients are used as features, and a subset of the Saak coefficients is selected from each stage. Next, the feature dimension is further reduced and fed into the SVM classifier for the classification task.

Experiments have been conducted on the MNIST dataset. Since the handwritten digits recognition problem is well solved by CNNs such as the LeNet-5, the performance of lossless and lossy Saak transform-based solutions has been compared comprehensively to that of LeNet-5 in terms of scalability, robustness, and efficiency in the original paper. In general, the lossless Saak transform achieved a classification accuracy of 98.54% and the lossy Saak transform achieved a classification accuracy of 98.53% on the MNIST dataset. The lossy Saak transform results in very less performance degradation, yet its complexity is significantly reduced.

4.1.3 Interpretable Convolutional Neural Networks via Feedforward Design

An interpretable feedforward (FF) design without any backpropagation (BP) [20] has been proposed in this work to obtain the model parameters. The FF design is a data-centric approach that derives the network parameters of the current layer based on data statistics from the output of the previous layer in a one-pass manner. In

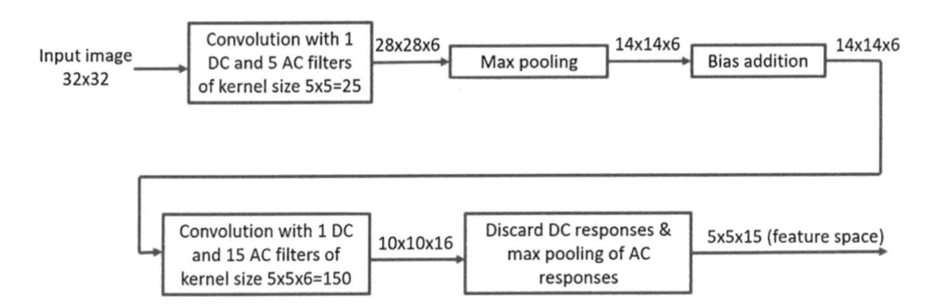

Fig. 4.3 Two-layer Saab transform. Reproduced with permission [20]. Copyright © 2019, Elsevier Inc.

our interpretation, each CNN layer corresponds to a vector space transformation. A sample distribution of the input space exists in the training data. To determine a proper transformation from the input to the output using the input data distribution, two steps are employed: (1) dimension reduction through subspace approximations and/or projections and (2) training sample clustering and remapping. The former is used in the construction of convolutional layers, while the latter is adopted to build fully connected (FC) layers.

The convolutional layers offer a sequence of spatial–spectral filtering operations. A new signal transform, called the Saab transform, has been developed to construct convolutional layers. This is a variant of the principal component analysis (PCA) with an added bias vector to annihilate the nonlinearity of the activation, and it contributes to dimension reduction. Multiple Saab transforms in cascade yield multiple convolutional layers (Fig. 4.3).

The FC layers provide a sequence of sample clustering and high- to low-dimensional mapping operations. It is constructed by a three-level hierarchy: the feature space, the sub-class space, and the class space. Linear least squared regression (LLSR) guided by pseudo-labels is adopted to build from the feature space to the sub-class space. Then, LLSR guided by true labels is used to build from the sub-class space to the class space. The FC layers formed by multistage LLSRs in cascade correspond to a multilayer perceptron (MLP). The design principle is not only to reduce the dimensions of the intermediate spaces but also to gradually increase the discriminability of some dimensions. The multilayer transformations eventually reach the output space with strong discriminability.

Generally speaking, the Saab transform is more advantageous than the Saak transform, because the number of Saab filters is only one half of the number of Saak filters. Besides interpreting the cascade of convolutional layers as a sequence of approximating spatial–spectral subspaces, the fully connected layers act as a sequence of label-guided least squares regression processes. As a result, all the model parameters of CNNs can be determined in an FF one-pass fashion. This is known as an FF-designed CNN (FF-CNN). The classification performances of BP and FF-designed CNNs and their robustness against adversarial attacks have been compared on the MNIST and the CIFAR-10 datasets in the original paper. In general, the FF-designed CNNs obtain a classification accuracy of 97.2% on the MNIST dataset and 62% on the CIFAR-10 dataset.

4.1.4 PixelHop

PixelHop [7] was the first method to introduce successive subspace learning (SSL) for feature extraction from 2D images. Even though the term SSL was first introduced here, the roots of SSL stem from the works discussed in the preceding subsections. PixelHop presents the SSL-based method for image classification task. It consists of four main steps: successive near-to-far neighborhood expansion, unsupervised dimension reduction via the Saab transform, supervised dimension reduction using label-assisted regression (LAG), and feature concatenation and classification. These operations form modules #1, #2, and #3 of the PixelHop system, which is illustrated in Fig. 4.4.

Module #1 consists of a sequence of PixelHop units in cascade. The PixelHop unit is the core step in feature construction and comprises neighborhood attribute construction and dimension reduction. The mechanism operates as follows. In the i-th PixelHop unit, the attributes of the target pixel and its N_i neighboring pixels, all of dimension $K_{(i-1)}$, are concatenated to obtain an enlarged neighborhood of size $K_{(i-1)} \times (N_i + 1)$. N_i is set to 8, representing the 8 neighbors of a pixel. In this manner, the attribute dimension becomes $K_0 \times 9^I$ after I PixelHop units. To control this sharp increase in dimension, it is desirable to reduce the number of dimensions after every neighborhood expansion step. Hence, dimension reduction is performed using the Saab transform [20]. After the i-th PixelHop unit, the spectral dimension is reduced from $9K_{i-1}$ to K_i due to the Saab transform, while the spatial resolution

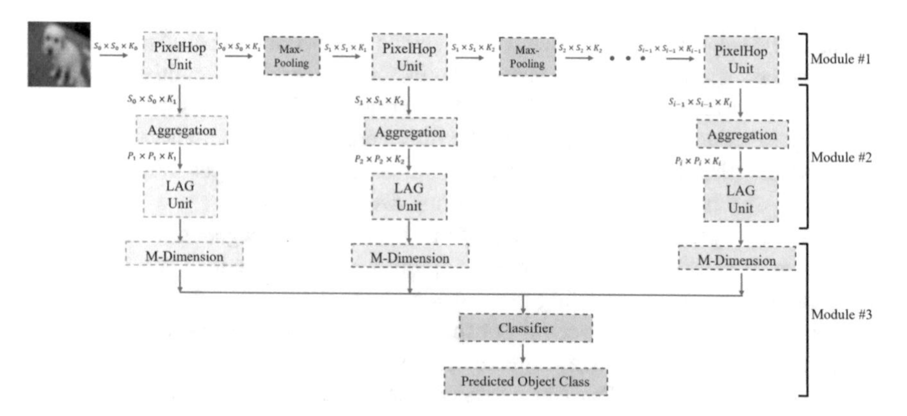

Fig. 4.4 System diagram of PixelHop. Reproduced with permission [7]. Copyright © 2020, Elsevier Inc.

remains the same. To account for spatial redundancy introduced by considering a stride of one, the (2×2)-to-(1×1) max pooling operation is performed between successive PixelHop units. Then, the spatial resolution is reduced from $S_{(i-1)} \times S_{(i-1)}$ to $S_i \times S_i$.

Module #2 consists of feature aggregation and supervised dimension reduction through label-assisted regression (LAG). To extract diverse features after every PixelHop unit, several feature aggregation techniques are considered, including maximum, minimum, and mean pooling of responses in small non-overlapping regions. Furthermore, the feature dimension is reduced using supervised learning. This acts as a bridge between the feature space and the decision space. For a given neighborhood size, the attributes of different object classes follow different distributions. The samples of every class are divided into a fixed number of clusters using k-means clustering. Clustering handles the intraclass variations among samples. The cluster centers are stored. Thereafter, a soft probability vector is obtained based on the distance between the feature vector of the target training image and the centroids of all clusters belonging to the target image class. Subsequently, a least squares regression model is employed to map the features to soft probability vectors. This operation is termed label-assisted regression (LAG), because class label is used to obtain class-wise clusters. For every hop, a separate LAG unit is used. Finally, the soft probability vectors from all LAG units are concatenated and a multiclass classifier is trained. This comprises the module #3 of PixelHop.

Experimentally, support vector machine (SVM) is used as a classifier with the radial basis function (RBF) kernel. PixelHop achieves a classification accuracy of 99.09% on the MNIST dataset and 91.68% on the Fashion MNIST, while the test accuracy on the CIFAR-10 dataset is 72.66%.

4.1.5 PixelHop++

PixelHop++ [8] is a follow-up work of PixelHop, with the following improvements. First, the model size of PixelHop++ is smaller than that of PixelHop. This is achieved by replacing the Saab transform with a new channel-wise Saab transform. The latter decouples a joint spatial–spectral input tensor into multiple spatial tensors for each spectral component and performs the Saab transform on each component separately. Furthermore, a subset of discriminant features is selected based on cross-entropy criteria to boost the classification performance. The system diagram of PixelHop++ is shown in Fig. 4.5. It is similar to that of PixelHop, except with the modified channel-wise Saab transform and an additional feature selection process.

The neighborhood construction step in module #1 of PixelHop++ is similar to that of PixelHop, so we do not repeat it here. The channel-wise Saab transform and the resulting tree-decomposed feature decomposition are constructed as follows. For the principal components analysis (PCA), all the output channels are uncorrelated. Because the Saab transform is derived from PCA, the Saab coefficients tend to be weakly correlated in the spectral domain. This hypothesis has been experimentally

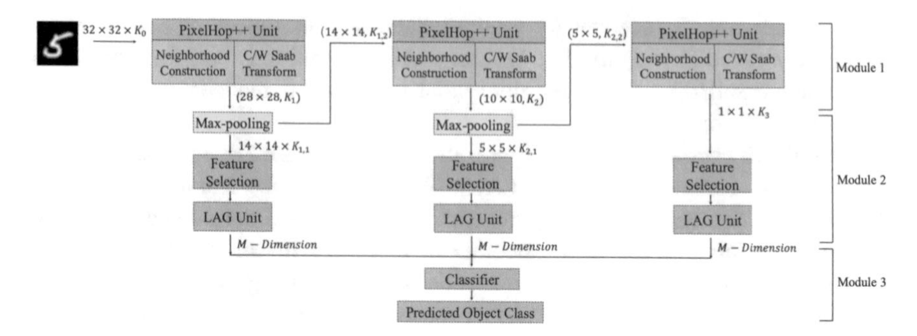

Fig. 4.5 System diagram of PixelHop++. Reproduced with permission [8]. Copyright © 2020, IEEE

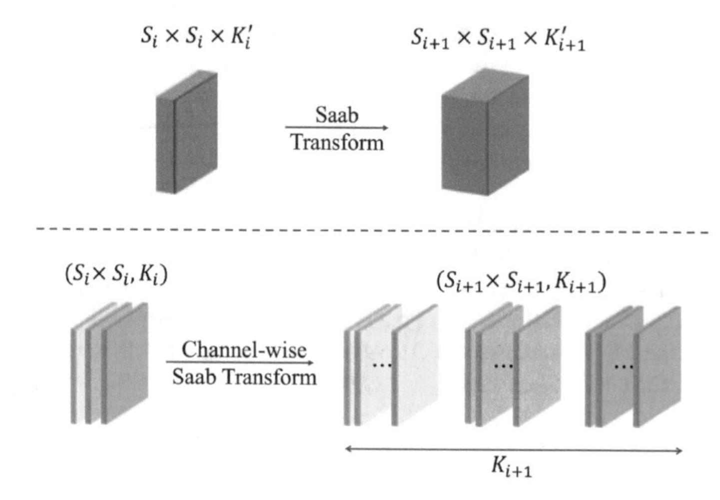

Fig. 4.6 Channel-wise (c/w) Saab transform. Reproduced with permission [8]. Copyright © 2020, IEEE

validated by observing the correlations among the output spectral components. The weak spectral correlations of Saab coefficients make it possible to decompose the joint spatial–spectral input tensor with spectral dimension of K_i in the i-th PixelHop++ unit into K_i spectral tensors of an appropriate spatial size. Then, separate Saab transforms are performed on each K_i spectral channel. This modified transform is thus called the channel-wise (c/w) Saab transform. A simple illustration that contrasts the two transforms is shown in Fig. 4.6. Like in PixelHop, a max pooling operation is performed on every channel output before proceeding to the next hop.

Multi-hop feature learning via successive neighborhood expansions coupled with the channel-wise Saab transform results in a feature tree representation. The root node of the tree is the input image. Every node in the feature tree is associated with an energy, and the root node is considered to be of unit energy. Each spectral

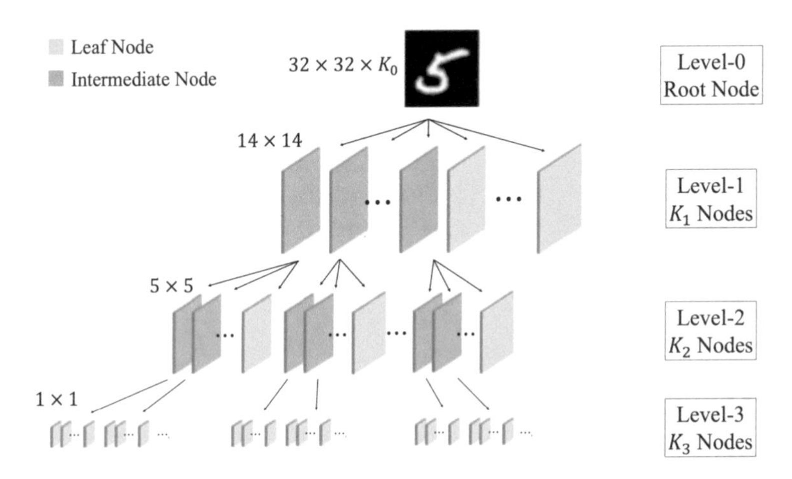

Fig. 4.7 Hierarchical tree decomposition. Reproduced with permission [8]. Copyright © 2020, IEEE

component at the output of the first PixelHop++ unit, followed by max pooling, represents one child node of the root. The energy of a child node is the product of the energy of its parent nodes and its normalized energy with respect to all its siblings. An energy threshold T is defined, which acts as a hyperparameter to control the growth of the feature tree. Child nodes with an energy of less than T are collected as leaf nodes. Since every node represents a single dimension output due to the channel-wise Saab transform, each leaf node contributes to a single feature dimension. The child nodes with energies greater than T are passed to the next hop. Such nodes are called intermediate nodes. An example feature tree is shown in Fig. 4.7. The leaf nodes at different levels in the feature tree have different spatial dimensions or receptive fields.

In the next step, the cross-entropy value for each feature at the leaf node is computed using the relation

$$L = \sum_{j=1}^{J} L_j, \quad L_j = -\sum_{c=1}^{M} y_{j,c} \log(p_{j,c}). \tag{4.1}$$

Here, M denotes the number of classes, $y_{j,c}$ is a binary indicator that shows whether sample j is correctly classified, and $p_{j,c}$ is the probability of sample j belonging to class c. The features are ordered in an increasing order of cross-entropy, and the top N_s features are selected. This feature set has lower cross-entropy and a higher discriminant power. The next operation of the LAG unit is similar to that in PixelHop. All the M features from every hop are concatenated and fed to a least squares regressor for final classification.

Similar to PixelHop, the performance of the PixelHop++ model has been experimentally tested using the MNIST, Fashion MNIST, and CIFAR-10 datasets.

The best PixelHop++ model achieves a classification accuracy of 98.49%, 90.17%, and 66.81% on these datasets, respectively, but with far fewer parameters than that used by PixelHop. The model size can be controlled by the energy threshold parameter T and the number of nodes selected after cross-entropy analysis N_s.

After reviewing the preliminary work on SSL, we now proceed to the SSL-based methods for point cloud processing. These methods are discussed in more depth in the following sections (Sects. 4.2 and 4.3). Later, we will come back to images and discuss some applications of SSL for 2D vision tasks other than image classification. These applications will use PixelHop++ as the backbone for feature extraction.

4.2 Classification and Part Segmentation

Traditional methods for point cloud classification and segmentation tasks usually use handcrafted feature descriptors, which tend to be geometric and/or shallow. Nevertheless, these methods do not need to be supervised, and they are quite efficient and readily interpretable. In contrast, deep learning methods require end-to-end supervision to complete tasks, and the learned features are more semantic owing to the high costs of computation resources like GPU. The high time and memory costs also make it challenging to deploy these methods in mobile or terminal devices. In addition, these methods are often criticized for the lack of interpretability. To address these problems, we propose explainable machine learning methods for point cloud classification and segmentation, which are data-driven like deep learning methods while learning features in a single feedforward pass like traditional methods. Explainable machine learning methods are mathematically transparent, which means that they are much faster and demand less memory. Moreover, their performance is on par with that of deep learning methods. PointHop [44] and PointHop++ [43] for point cloud classification and unsupervised feedforward feature (UFF) learning for point cloud classification and segmentation [42] are introduced in this section.

PointHop is the first explainable machine learning method for point cloud data recognition. Its process flow is compared with that of deep learning methods in Fig. 4.8. Other proposed explainable machine learning methods share the same characteristics as PointHop. For deep learning methods, point cloud data are fed into DNNs in the feedforward pass, and then the losses and gradients are calculated and propagated in the backward direction to update the parameters. This process is conducted iteratively until convergence. Labels are needed to update all model parameters. In contrast, for explainable machine learning methods, point cloud data are fed into the self-designed system, e.g., PointHop, to build and extract features in a single fully explainable feedforward pass. No labels are needed in the feature extraction stage (i.e., unsupervised feature learning). The whole training of PointHop can be efficiently performed using a single CPU, as its complexity is much lower than that of deep-learning-based methods.

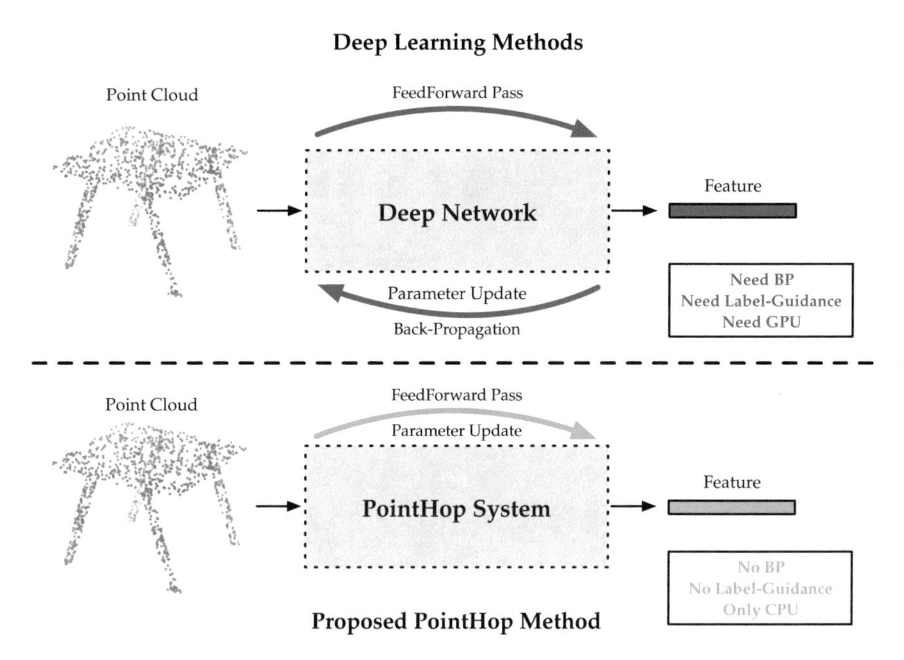

Fig. 4.8 Comparison of deep learning methods and the proposed PointHop method. Reproduced with permission [44]. Copyright © 2020, IEEE

4.2.1 *PointHop*

A point cloud of N points is defined as $\mathbf{P} = \{\mathbf{p}_1, \cdots, \mathbf{p}_N\}$, where $\mathbf{p}_n \in \mathbb{R}^3$, $n = 1, \cdots, N$. There are two distinct properties of the point cloud data:

- Unordered data in the 3D space
  Point clouds comprise a set of points in the 3D space without a specific order, which is different from images where pixels are defined in a regular 2D grid.
- Disturbance in scanned points
  For the same 3D object, different point sets can be acquired with uncertain position disturbance. For example, different scanning methods are applied to the surface of the same object or the scanning device is at different locations.

An overview of the proposed PointHop method is shown in Fig. 4.9. A point cloud, **P**, is taken as the input, and PointHop outputs the corresponding class label. The PointHop mechanism consists of two stages: (1) local-to-global attribute building through multi-hop information exchange and 2) classification and ensembles. The input point cloud has N points with three coordinates (x, y, z). It is fed into multiple PointHop units in cascade and their outputs are aggregated by M different schemes to derive features. All features are cascaded for object classification.

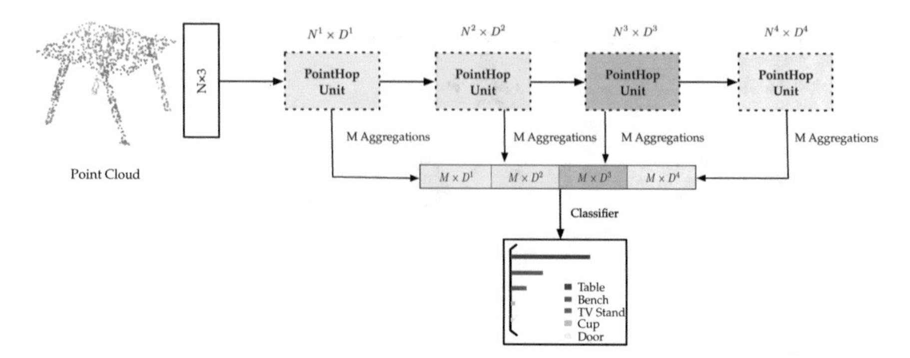

Fig. 4.9 Overview of the PointHop method. Reproduced with permission [44]. Copyright © 2020, IEEE

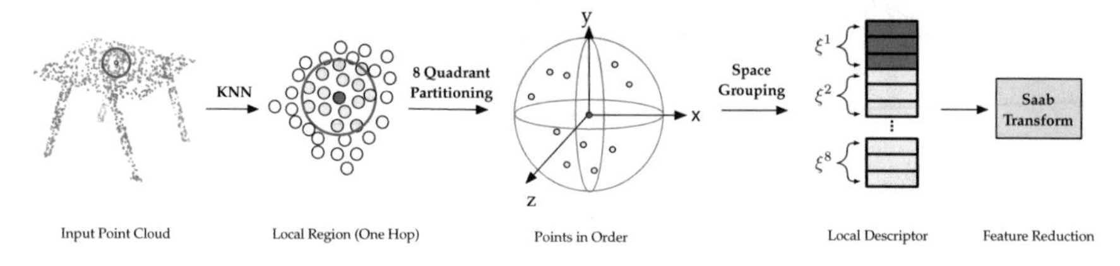

Fig. 4.10 Illustration of the PointHop unit. The red point is the center point, while the yellow points represent its K nearest neighbors. Reproduced with permission [44]. Copyright © 2020, IEEE

Local-to-global Attribute Building The attribute building stage addresses the problem of unordered point cloud data using a space partitioning procedure and developing a robust descriptor that characterizes the relationship between a point and its one-hop neighbors. Initially, the attributes of a point are its 3D coordinates. When multiple PointHop units are performed in cascade, the attributes of a point will grow by considering its relationship with one-hop neighbor points iteratively. The attributes of a point evolve from a low-dimensional vector into a high-dimensional vector through this module. To control this rapid dimension growth, the Saab transform is applied for dimension reduction, so that the dimension grows at a slower rate. All these operations are conducted inside the PointHop processing unit. The local descriptor is robust since the construction acts to minimize the issues of unordered 3D data and disturbance of scanned points into account.

The PointHop unit is shown in Fig. 4.10. For each point in the point cloud $\mathbf{P}$, we search its K nearest neighbor points for $\mathbf{p}_c = (x_c, y_c, z_c)$ in the point cloud, including the point itself, where the distance is measured by the Euclidean norm. The point and its K nearest neighbors form a local region:

$$KNN(\mathbf{p}_c) = \{\mathbf{p}_{c1}, \cdots, \mathbf{p}_{cK}\}, \quad \mathbf{p}_{c1}, \cdots, \mathbf{p}_{cK} \in \mathbf{P}. \tag{4.2}$$

We treat $\mathbf{p}_c$ as a new origin for each local region centered at $\mathbf{p}_c$, so that we can partition the local region into eight octants ξ^j, $j = 1, \cdots, 8$, based on the value of each coordinate (i.e., greater or less than that of $\mathbf{p}_c$).

The centroid of the point attributes in each octant is computed via

$$\mathbf{a}_c^j = \frac{1}{K_j} \sum_{i=1}^{K_j} t_{ci}^j \mathbf{a}_{ci}, \quad j = 1, \cdots, 8, \tag{4.3}$$

where $\mathbf{a}_{ci}$ is the attribute vector of point $\mathbf{p}_{ci}$ and

$$t_{ci}^j = \begin{cases} 1, & x_{ci} \in \xi^j, \\ 0, & x_{ci} \notin \xi^j \end{cases} \tag{4.4}$$

is the coefficient to indicate whether point x_{ci} is in octant ξ^j and K_j is the number of K-NN points in octant ξ^j. Finally, all centroids of attributes $\mathbf{a}_c^j$, $j = 1, \cdots, 8$, are concatenated to form a new descriptor of sampled point $\mathbf{p}_c$:

$$\mathbf{a}_c = Concat\{\mathbf{a}_c^j\}_{j=1}^8. \tag{4.5}$$

This descriptor is robust with respect to disturbance in the positions of the acquired points because of the averaging operation in each octant.

Definition 4.1 If $\mathbf{p}_C$ is a 1-hop neighbor of $\mathbf{p}_B$ and $\mathbf{p}_B$ is a 1-hop neighbor of $\mathbf{p}_A$ while $\mathbf{p}_C$ is not a 1-hop neighbor of $\mathbf{p}_A$, we call $\mathbf{p}_C$ is a 2-hop neighbor of $\mathbf{p}_A$. If $\mathbf{p}_C$ is a 1-hop neighbor of $\mathbf{p}_B$ and $\mathbf{p}_B$ is a 1-hop neighbor of $\mathbf{p}_A$, we call $\mathbf{p}_C$ is a 2-hop neighbor of $\mathbf{p}_A$ if $\mathbf{p}_C$ is not a 1-hop neighbor of $\mathbf{p}_A$.

The initial attributes of a point are known as 0-hop attributes. Where the initial attributes are the 3D coordinates (x, y, z) of a point, the dimension of 0-hop attributes is 3. Note that the 0-hop attributes can be generalized to (x, y, z, r, g, b) for point clouds with color information (r, g, b) at each point. The local descriptor, as given in Eq. 4.5, has a dimension of $3 \times 8 = 24$. We adopt the local descriptor as the new attributes of a point that considers its relationship with its K-NN neighbors. These are known as the 1-hop attributes. The n-hop attributes characterize the relationship of a point with its m-hop neighbors, $m \leq n$. As n becomes larger, the n-hop attributes offer a larger coverage of points in a point cloud model, which is analogous to a larger receptive field in deeper layers of CNNs.

The dimension of the attribute vector of each point grows from 3 to 24 due to the change of local descriptors from 0-hop to 1-hop. We then build another local descriptor based on the 1-hop attributes of each point. This descriptor defines the 2-hop attributes of dimension $24 \times 8 = 192$. The attribute dimension grows fast. It is desirable to reduce the dimension of the n-hop attribute vector first before reaching out to neighbors of the $(n + 1)$-hop. The Saab transform [20] is used to reduce the attribute dimension of each point. Each PointHop unit includes a one-stage Saab

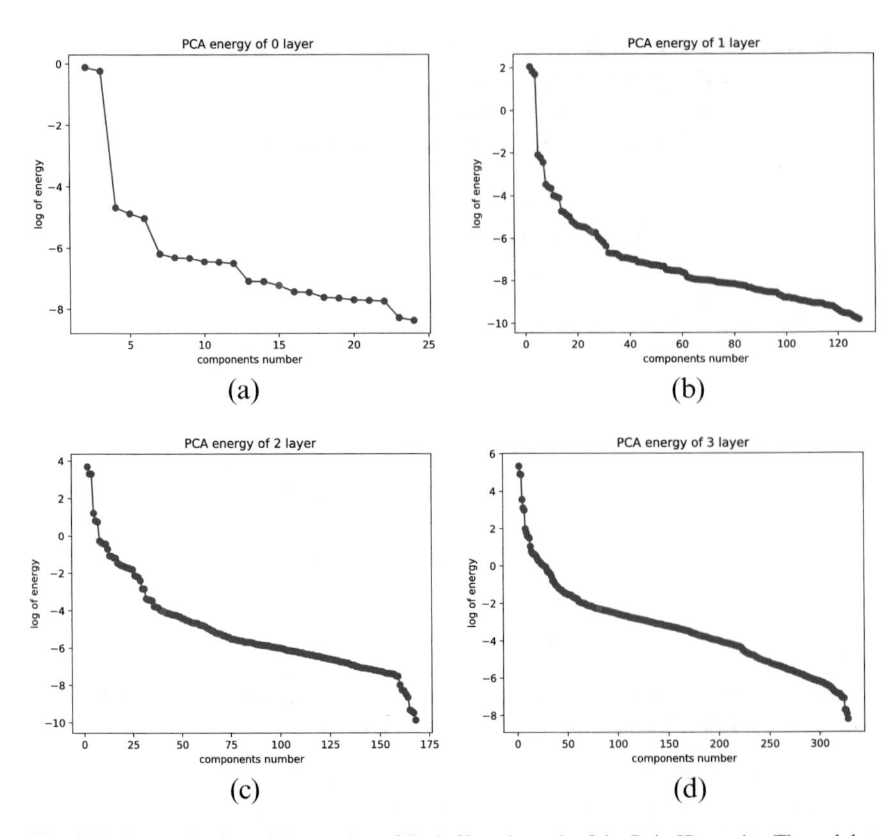

Fig. 4.11 Determination of the number of Saab filters in each of the PointHop units. The red dots indicates the selected number of Saab filters. Reproduced with permission [44]. Copyright © 2020, IEEE. (**a**) First unit. (**b**) Second unit. (**c**) Third unit. (**d**) Fourth unit

transform. For L PointHop units in cascade, we need L-stage Saab transforms. We set $L = 4$ in the experiments. To determine the number of Saab filters required, energy plots of the PCA coefficients are drawn for each PointHop unit, as shown in Fig. 4.11. The knee location of the curve, as indicated by the red point in each subfigure, is used to identify the appropriate number of Saab filters.

To reduce the computational complexity and increase the coverage rate, a spatial sampling scheme is adopted between two consecutive PointHop units, so that the number of points to be processed is reduced. This is achieved by using the farthest point sampling (FPS) scheme [12, 16, 24], which captures the geometrical structure of a point cloud model well.

For each PointHop unit, we aggregate (or pool) the points' features into a single global feature vector. To enrich the feature set, we consider multiple aggregation/pooling schemes such as the max pooling [25], the mean aggregation, the l_1-norm aggregation, and the l_2-norm aggregation. Then, we concatenate them to obtain a feature vector of dimension $M \times D^i$, where M is the number of attribute aggregation methods, and D^i is the dimension of feature kept for the i-th PointHop unit. Finally, we concatenate the feature vectors of all PointHop units to form an ultimate feature vector of the whole system.

Classification and Ensembles Well-known classifiers such as support vector machine (SVM) [10] and random forest (RF) [4] classifiers are adopted to perform the classification task with the extracted features. The SVM classifier performs classification by finding the gaps that separate different classes. Test samples are then mapped into one of the sides of the gap and predicted to be the label of that side. The RF classifier first trains a number of decision trees, each of which provides an output. Then, the outputs of all decision trees are ensemble to give the mean prediction. Both classifiers are mature and easy to use.

Ensembles are adopted to further improve the classification performance. We adopt the feature ensemble strategy which cascades features by different schemes and offers better classification accuracy at the cost of a higher complexity if the feature dimension is high. With the feature ensemble strategy, it is desirable to increase the diversity of PointHop to enrich the feature set. We use the following four schemes to achieve this goal. First, we augment the input data by rotating it by a certain degree. Second, we change the number of Saab filters in each PointHop unit. Third, we change the K value in the K-NN scheme. Fourth, we vary the numbers of points in each PointHop unit.

Experiments Experiments were conducted on the ModelNet40 [37], a popular 3D object classification dataset. This dataset contains 40 categories of CAD models of objects such as airplanes, chairs, benches, cups, and so on. There are 9843 training samples and 2468 testing samples. Each initial point cloud has 2048 points and each point has three Cartesian coordinates. Following the settings of PointNet [25], we downsampled the initial point cloud to 1024 points.

The classification accuracy for different pooling methods is shown in Fig. 4.12, as a function of the number of sampled points of the input model. The ensemble of all pooling schemes (the red line) gives the best results regardless of the number of sampled points selected. In addition, the highest accuracy is achieved when we use 768 or 1024 points are used with the ensemble of all four pooling schemes.

The classification accuracy of the proposed PointHop system with the 1024-point model is compared with those of several state-of-the-art methods in Table 4.1. The column of "average accuracy" means the average of per-class classification accuracy, while the column of "overall accuracy" shows the accuracy of the best result obtained. Our PointHop baseline, which contains a single model without any ensembles, achieves an overall accuracy of 88.65%, which is better than that of some unsupervised methods such as LFD-GAN [1] and FoldingNet [39]. With ensemble, the overall accuracy can be further increased to 89.1%, which is only 0.1% and 3.1% worse than those of PointNet [22] and DGCNN [35], respectively.

Next, the time complexity is compared in Table 4.2. The training time of the PointHop system is significantly lower than that of the deep-learning-based methods. The PointHop system was training using an Intel(R) Xeon(R) CPU E5-

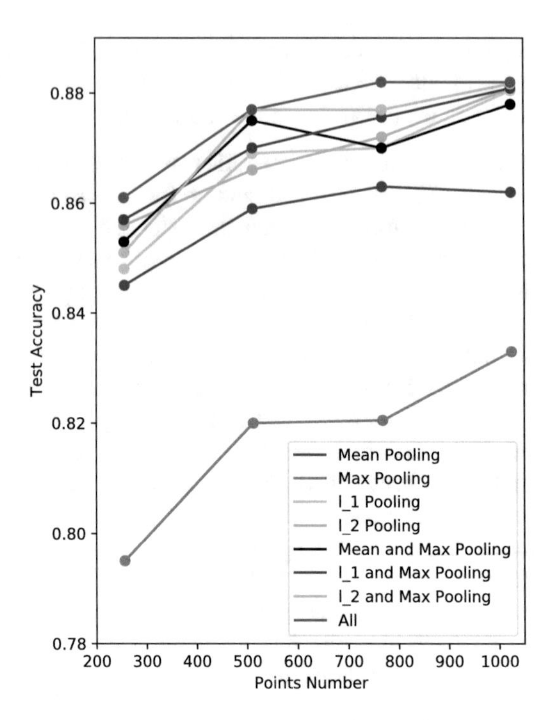

Fig. 4.12 Feature aggregation. Reproduced with permission [44]. Copyright © 2020, IEEE

Table 4.1 Comparison of classification accuracy on ModelNet40. Reproduced with permission [44]. Copyright © 2020, IEEE

Method	Feature extraction	Average accuracy (%)	Overall accuracy (%)
PointNet [25]	Supervised	86.2	89.2
PointNet++ [26]		–	90.7
PointCNN [22]		88.1	92.2
DGCNN [35]		90.2	92.2
PointNet baseline (Handcraft, MLP)	Unsupervised	72.6	77.4
LFD-GAN [1]		–	85.7
FoldingNet [39]		–	88.4
PointHop (baseline)		83.3	88.65
PointHop		84.4	89.1

Table 4.2 Comparison of time complexity. Reproduced with permission [44]. Copyright © 2020, IEEE

Method	Total training time	Inference time (ms)	Device
PointNet (1024 points)	~5 h	25.3	GPU
PointNet++ (1024 points)	–	163.2	GPU
PointHop (256 points)	~5 min	103	CPU
PointHop (1024 points)	~20 min	108.4	CPU

2620 v3 at 2.40 GHz. The PointHop baselines with 256- and 1024-point cloud models were trained in 5 and 20 min, respectively. In contrast, PointNet [25] takes more than 5 h to train using one GTX1080 GPU. Furthermore, we compare the inference time in the test stage. PointNet++ takes 163.2 ms to classify a test sample of 1024 points with GPU, whereas the PointHop method only needs 108.4 ms of CPU.

4.2.2 PointHop++

The PointHop method shows good classification performance on the ModelNet40 dataset with low training complexity. However, there are two shortcomings of the PointHop method. First, while each point has a small receptive field in the beginning of the pipeline, the receptive field increases in size further into the pipeline. The PointHop system trades a larger spatial dimension for a higher spectral dimension. We use $n_t = n_a \times n_e$ to denote the tensor dimension at a certain hop, where n_a and n_e are the spatial and spectral dimensions, respectively. Under the SSL framework, we need to compute the covariance matrix of vectorized tensors, which has a dimension of $n_t \times n_t$. Then, the complexity is $O(dn_t^2 + d^3)$ to find d principal components. Since $n_t > d$, the first term dominates. To reduce the size of the learning model, it is desirable to lower the input tensor dimension so as to reduce the filter size. The second shortcoming is that PointHop does not include a loss function minimization, yet this functionality plays an important role in deep-learning-based methods.

A tree-structured unsupervised feature learning system named the PointHop++ method is proposed to improve the PointHop method furthermore from two aspects: (1) to reduce the model complexity in terms of the model parameter number and (2) to automatically order discriminant features based on the cross-entropy criterion and perform feature selection for better performance. The first improvement is essential for wearable and mobile computing, while the second improvement bridges statistics-based and optimization-based machine learning methodologies.

An overview of the proposed PointHop++ method is illustrated in Fig. 4.13. A point cloud set, **P**, which consists of N points denoted by $p_n = (x_n, y_n, z_n)$, $1 \leq n \leq N$, is taken as the input of the feature learning system to obtain a powerful feature representation. N^i denotes the number of points in the input and in the n-th hop. The upper-left enclosed subfigure shows the operation in the first PointHop unit. We decouple the joint spatial–spectral tensor into single spectral components, because the correlation between different spectral channels is very weak. This channel-wise (c/w) subspace decomposition helps to reduce the model complexity of PointHop with regard to the number of model parameters and required computational memory. Subspaces with energies larger than a threshold energy T proceed to the next hop, while the other subspaces become leaf nodes of the feature tree in the current hop. Through multiple decomposition stages, we obtain a one-dimensional (1D) feature at each leaf node of the feature tree. Then, we use the cross-entropy loss function to rank features so that we can select a subset of

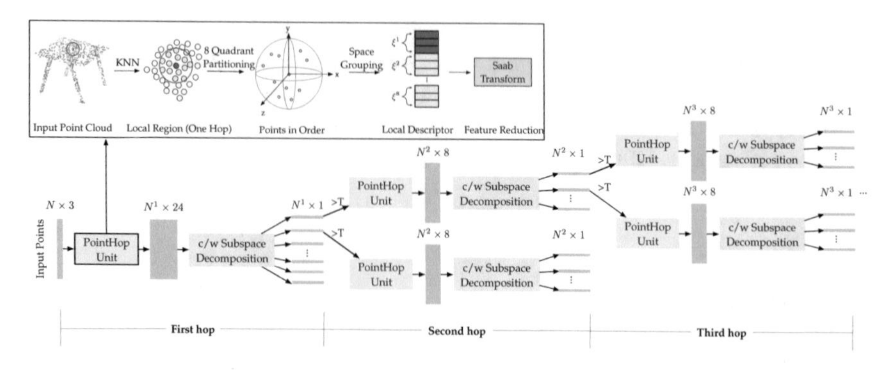

Fig. 4.13 Overview of the PointHop++ method. Reproduced with permission [43]. Copyright © 2020, IEEE

discriminant features to train classifiers. Finally, the linear least squares regression (LLSR) is conducted on the obtained features to output a 40-dimensional probability vector, from which the corresponding class labels are selected.

Initial Feature Space Construction To extract the local features of each point $p_c \in P$, we follow the same design principle as that of the PointHop unit. The k nearest neighboring points of point p_c are retrieved to build a neighboring point set that includes p_c itself. The neighborhood set is partitioned into eight octants according to their relative spatial coordinates. Then, the mean pooling is applied to the attributes in each octant. Mathematically, we have the following mapping:

$$g : \underbrace{\mathbb{R}^D \times \cdots \mathbb{R}^D}_{\text{k}} \rightarrow \underbrace{\mathbb{R}^D \times \cdots \mathbb{R}^D}_{8}, \tag{4.6}$$

where $D = 3$ for the first hop and $D = 1$ for the subsequent hops. For the first hop, we use the spatial coordinates $p_n = (x_n, y_n, z_n)$ as the attributes. For the subsequent hops, we use a 1D spectral component as the attribute of retrieved points because of the usage of c/w subspace decomposition to the output from the previous hop. Attributes of all eight octants are concatenated to become $\mathbf{a} \in \mathbb{R}^{8D}$, which represents the attributes of selected point p_c before c/w subspace decomposition.

Channel-Wise Subspace Decomposition PointHop uses the Saab transform [20] as a dimension reduction tool, which is a variant of the PCA [36] designed to overcome the sign confusion problem [18] when multiple PCA stages are in cascade. The Saab transform is also used in PointHop++. However, where the Saab transform coefficients are all grouped together and input to the next hop unit in PointHop, they are decomposed into 1D subspaces that represent a spatial–spectral localized representation of the point set and are input to the next hop units independently in PointHop++. To decompose the eight-dimensional (8D) Saab coefficient vector into eight 1D subspaces, we must first prove that the

Saab coefficients of different channels are weakly correlated. Besides the physical meaning, this representation demands less computation in the next hop.

The correlation of Saab coefficients is computed to validate c/w subspace decomposition. The input to the Saab transform is

$$A = [\mathbf{a}^1, \cdots, \mathbf{a}^N]^T \in \mathbb{R}^{N \times 8D},$$

where $\mathbf{a}^n$ is the 8D attribute vector of point p_n, and the filter weight is

$$W = [\mathbf{w}_1, \mathbf{w}_2, \cdots, \mathbf{w}_{8D}] \in \mathbb{R}^{8D \times 8D},$$

where $\mathbf{w}_1 = \frac{1}{\sqrt{8D}}[1, 1, \cdots, 1]^T$ and the other filter weights are eigenvectors of covariance matrix of A ranked by their associated eigenvalue λ_i decreasingly. The output of the Saab transform is

$$B = A \cdot W = [\mathbf{b}_1, \cdots, \mathbf{b}_{8D}],$$

where $\mathbf{b}_i \in \mathbb{R}^{N \times 1}, i = 1, \cdots, 8D$. Hence, the correlation between Saab coefficients of different channels is

$$Cor(\mathbf{b_i}, \mathbf{b_j}) = \frac{1}{N}(A \cdot \mathbf{w}_i)^T (A \cdot \mathbf{w}_j) = \frac{1}{N}(\lambda_i \mathbf{w}_i)^T (\lambda_j \mathbf{w}_j) = 0, \tag{4.7}$$

where $i > 1$, $j > 1$, and $i \neq j$. The last equality comes from the orthogonality of eigenvectors in the PCA analysis (Fig. 4.14).

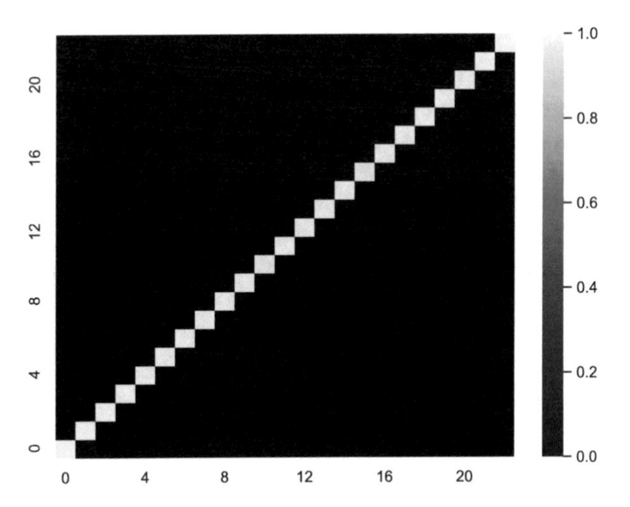

Fig. 4.14 Visualization of the correlation matrix at the first hop for $\mathbf{w}_i$, $i = 2, \cdots, 8D$. Reproduced with permission [43]. Copyright © 2020, IEEE

This justifies the decomposition of a joint feature space into multiple uncorrelated 1D subspaces as

$$\mathbb{R}^{8D} \rightarrow \underbrace{\mathbb{R}^1 \times \cdots \mathbb{R}^1}_{8\text{D}}. \tag{4.8}$$

In practice, we observe a very weak correlation between Saab coefficients (with a magnitude of 10^{-4}) compared to the diagonal term (i.e., self-correlation) because of the special choice of the first filter weight $\mathbf{w}_1$. The above analysis is only an approximation.

Channel Split Termination and Feature Priority Ordering The energy of each subspace is calculated as

$$E_i = E_p \times \frac{\lambda_i}{\sum_{j=1}^{8D} \lambda_j}, \tag{4.9}$$

where $i = 1, \cdots, 8D$ and E_p is the energy of its parent node. We preset a universal threshold energy T for the whole system. If the energy of a node is less than the threshold T, we terminate its further split and keep it as a leaf node of the feature tree at the current hop. Otherwise, it will proceed to the next hop. After finishing the feature tree construction, all leaf nodes are collected as the feature representation.

To determine the threshold energy T, the training and validation accuracy curves are plotted as a function of $-\log(T)$ in Fig. 4.15a. Here, the training accuracy increases monotonously when T decreases from 0.1 to 0.00001 (equivalent to an increase of $-\log(T)$ from 1.0 to 5.0). However, the overall validation accuracy peaks at $T = 0.0001$ ($-\log(T) = 4.0$) with a value of 90.3%, it rapidly decreases as T becomes smaller. Thus, $T = 0.0001$ is selected as the universal threshold.

After collecting all leaf nodes as the feature representation, the features are then ordered and selected based on their discriminant power to avoid overfitting. Since a feature is more discriminant if its cross-entropy is lower, the cross-entropy is computed for each feature at a leaf node. We follow the same process that described in [20]. First, the 1D subspace is partitioned into J intervals by a clustering algorithm [32]. Then, the majority vote is used to predict the label for each interval. With the ground truth labels, the probability that each sample belongs to a class can be obtained. Mathematically, we have

$$L = \sum_{j=1}^{J} L_j, \quad L_j = -\sum_{c=1}^{M} y_{j,c} \log(p_{j,c}), \tag{4.10}$$

where M is the class number, $y_{j,c}$ is a binary indicator to show whether sample j is correctly classified, and $p_{j,c}$ is the probability that sample j belongs to class c.

The training and validation accuracy curves with respect to the number of selected top-ranked features are plotted in Fig. 4.15b. The cross-entropy and energy

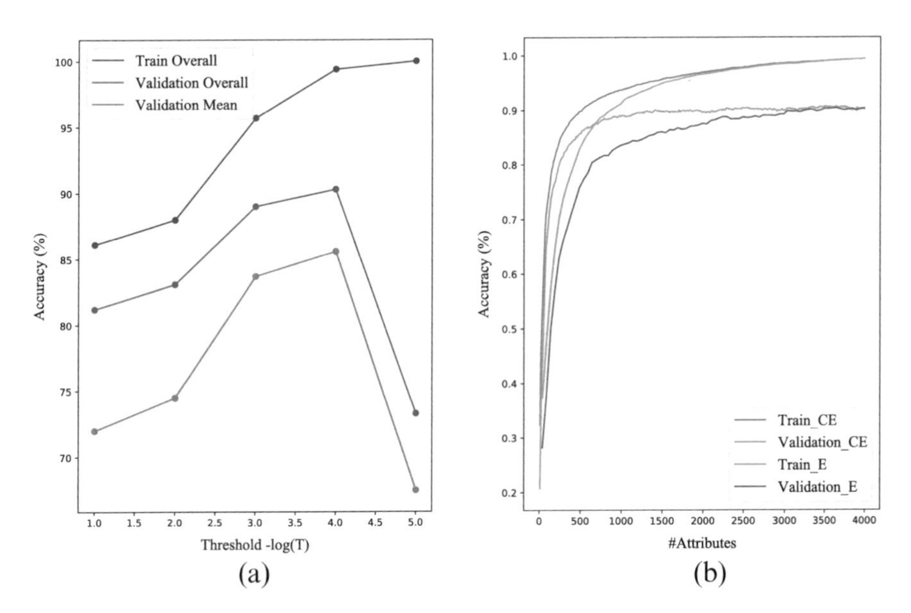

Fig. 4.15 Illustration of the impact of (**a**) values of the energy threshold T and (**b**) the number of cross-entropy-ranked (CE) or energy-ranked (E) features. Reproduced with permission [43]. Copyright © 2020, IEEE

ranking methods are compared. We see that overfitting is alleviated by both methods, while the cross-entropy ranking method performs better when the total feature number is smaller.

Experiments Experiments were conducted on the ModelNet40 dataset [37], following the same data processing and experimental settings as those used for the PointHop method. The depth of the feature tree was set to four hops, and the energy threshold T was set to be 0.0001. The classification accuracy is shown in Table 4.3. PointHop++ (baseline) (without feature selection (FS) or ensembles (ES)) has an overall accuracy of 90.3% and a class-average (class-avg) accuracy of 85.6%. The incorporation of the cross-entropy feature selection method (PointHop++ (FS)) improves the overall and class-avg accuracy results by 0.5% and 0.9%, respectively. An ensemble method is adopted to further improve the overall accuracy to 91.1% and the class-avg accuracy to 87%. Specifically, we rotate point clouds by 45 degrees and get a 40D feature by LLSR at each step. These features are then concatenated and fed into another LLSR. Compared with other unsupervised methods, PointHop++ method achieves the best performance. Particularly, it outperforms our previous work PointHop [44] by 2% in overall accuracy. Compared with deep networks, PointHop++ outperforms PointNet [25] and PointNet++ [26].

A comparison of the time complexity and model sizes of different methods is given in Table 4.4. PointHop++ took 25 min to train on an Intel(R)Xeon(R) CPU for a 1024-point cloud model, while the three deep networks took at least 7 h to train

Table 4.3 Comparison of classification results on ModelNet40. Reproduced with permission [43]. Copyright © 2020, IEEE

	Method	Accuracy (%)	
		Class-avg	Overall
Supervised	PointNet [25]	86.2	89.2
	PointNet++ [26]	–	90.7
	PointCNN [22]	88.1	92.2
	DGCNN [35]	90.2	92.2
Unsupervised	LFD-GAN [1]	–	85.7
	FoldingNet [39]	–	88.4
	PointHop [44]	84.4	89.1
	PointHop++ (baseline)	85.6	90.3
	PointHop++ (FS)	86.5	90.8
	PointHop++ (FS+ES)	87	91.1

Table 4.4 Comparison of time and model complexity. Reproduced with permission [43]. Copyright © 2020, IEEE

Method	Time		Parameter no. (MB)		
	Training (h)	Inference (ms)	Filter	Classifier	Total
PointNet [25]	7	10	–	–	3.48
PointNet++ [26]	7	14	–	–	1.48
DGCNN [35]	21	154	–	–	1.84
PointHop [44]	0.33	108	0.037	–	–
PointHop++	0.42	97	0.009	0.15	0.159

on a single GeForce GTX TITAN X GPU. For the inference time of each sample, both PointHop and PointHop++ took about 100 ms on a CPU. The number of model parameters was also calculated to show space complexity. The Saab filter size of PointHop++ was four times lower for PointHop++ than for PointHop. In addition, the total model parameters of PointHop++ were twenty times less than those of PointNet [25] and ten times less than those of DGCNN [35].

We further compare the robustness of different models against sampling density variation in Fig. 4.16. All the models were trained on a 1024-point cloud model and the point cloud was randomly downsampled to 768, 512, or 256 points, respectively, in the test stage. PointHop++ shows more robustness under mismatched sampling densities than PointHop [44], PointNet++ with single scale grouping (SSG) [26], and DGCNN [35].

4.2.3 *Unsupervised Feedforward Feature (UFF)*

PointHop and PointHop++ focus only on solving the point cloud classification problem. Therefore, we proposed a new solution for joint point cloud classification

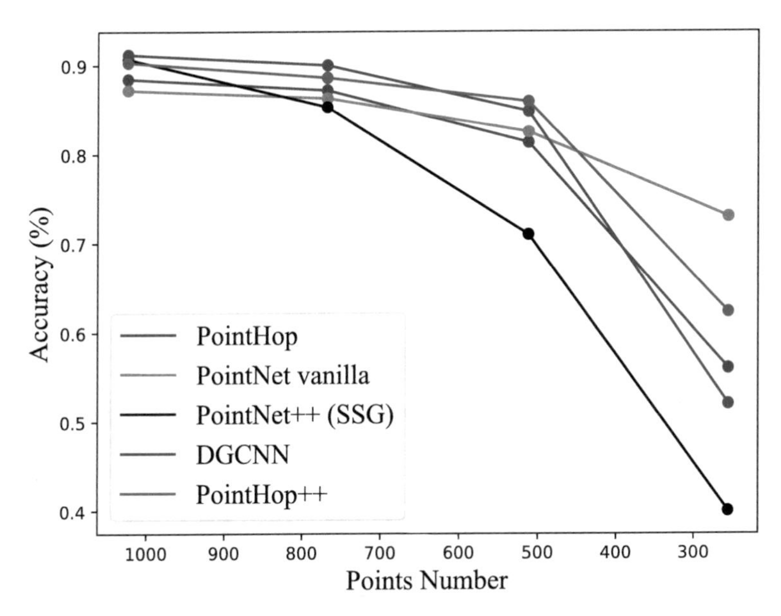

Fig. 4.16 Robustness of different test model against different sampling densities. Reproduced with permission [43]. Copyright © 2020, IEEE

and part segmentation here by generalizing the PointHop method. The main contribution of the new solution is the development of an unsupervised feedforward feature (UFF) learning system with fine-to-coarse (F2C) encoder and coarse-to-fine (C2F) decoder in cascade. Such an encoder–decoder architecture is frequently used in image segmentation. UFF exploits statistical correlation between points in a point cloud set to learn the parameters of the encoder and the decoder in a one-pass feedforward manner. Since it is a statistics-centric (rather than optimization-centric) approach, no labeling (or iterative optimization via backpropagation) is needed.

An overview of the UFF system is given in Fig. 4.17. It takes a point cloud as input and generates its shape and point features as the output. The global shape features are generated by the encoder for point cloud classification. Part segmentation is typically formulated as a pointwise classification task. For part segmentation, discriminant features must be found for all points in the original input point cloud. The downsampled resolution has to be interpolated back to a finer resolution in a layer-by-layer manner. The spatial coordinates and attributes of points at each layer are recorded by the encoder. Then, the decoder is used to generate new attributes of points layer by layer in a backward fashion. Finally, the extracted shape and point features of the input point cloud are fed to classifiers for shape classification and part segmentation.

Encoder Architecture The encoder follows the design principle of the PointHop method [44]. It has four layers, where each layer is a PointHop unit. A PointHop unit is used to summarize the information of a center pixel and its neighbor pixels.

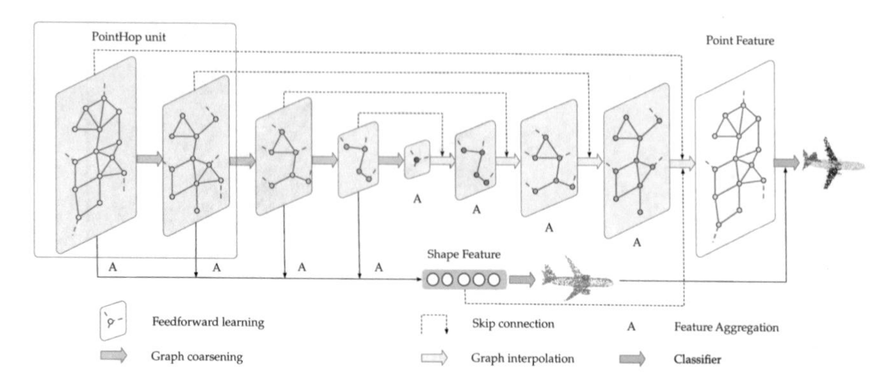

Fig. 4.17 Overview of the proposed UFF learning system. Reproduced with permission [42]. Copyright © 2020, IEEE

The PointHop units from the first to the fourth layers summarize the structures of 3D neighborhoods at short-, mid-, and long-range distances, respectively. The main operations of the PointHop units include (1) local neighborhood construction by the nearest neighbor search algorithm, (2) eight-octant 3D spatial partitioning and averaging of point attributes from the previous layer in each octant for feature extraction, and (3) feedforward convolution using the Saab transform for dimension reduction. Between two consecutive PointHop units, a point pooling operation is adopted based on the farthest point sampling (FPS) principle. We reduce the sampled points of a point cloud and enlarge the receptive field from one layer to the next by applying the FPS iteratively.

Decoder Architecture A decoder is used to obtain the features of points at the $(l-1)$-th layer based on the point features at the l-th layer. Its operations are similar to those of the encoder, with minor modifications. For every point at the $(l-1)$-th layer, we perform the nearest neighbor search to find its neighbor points located at the l-th layer. Then, eight-octant spatial partitioning and averaging of point attributes at the l-th layer are conducted in each octant for feature extraction. Next, we aggregate the features from each octant. Finally, we perform the feedforward convolution using the Saab transform for dimension reduction. It is worth emphasizing the difference between our decoder and that of PointNet++ [26]; that is, the latter calculates the weighted sum of the features of the neighbors according to their normalized spatial distances.

Roles of the Encoder/Decoder It is important to emphasize the difference between the attributes of a point at the encoder and at the decoder. At the encoder, the attribute vector of a point is constructed using a bottom-up approach, without a global view of the earlier layers. On the other hand, at the decoder, the attribute vector of a point at the decoder is constructed using a bottom-up approach followed by a top-down approach, which automatically includes global information. For convenience, we order the layers of the decoder backward; that is, the innermost layer is layer 4, followed by layers 3, 2, and 1. The outputs of the corresponding layers (with the same scale) between the encoder and the decoder are skip-connected, as shown in Fig. 4.17.

Feature Aggregation Feature aggregation was introduced in [44] to reduce the dimension of a feature vector while preserving its representation power. Four aggregation schemes were adopted ($M = 4$), including the mean, l_1-norm, l_2-norm, and l_∞-norm (i.e., max pooling) of the input vector. For a D-dimensional vector $\mathbf{a} = (a_1, \cdots, a_D)^T$, the key information is extracted to M aggregated values, where $M < D$. Then, we can reduce the dimension of the vector from D to M. We apply the same feature aggregation scheme here. Feature aggregation is denoted by A in Fig. 4.17.

Let N^l and D^l be the point number and the attribute dimension per point at the l-th layer, respectively. For the encoder, the raw feature map of a point cloud is a 2D tensor of dimension $N^l \times D^l$. Feature aggregation is conducted along the point dimension to obtain the shape feature; therefore, the aggregated feature map is a 2D tensor of dimension $M \times D^l$, where $M = 4$ is the number of aggregation methods. For the decoder, feature aggregation is conducted on points of the local neighboring region. The raw feature map at each point after eight-octant partitioning grouping is a 2D tensor of dimension $S \times D^{l-1}$, where $S = 8$ is the number of octants. Feature aggregation is conducted along the S dimension, and the aggregated feature map is a 2D tensor of dimension $M \times D^{l-1}$ with $M = 4$.

Integration with Classifiers For point cloud classification, the responses from all layers of the encoder are concatenated and aggregated as shape features. They are then fed into a classifier to obtain object labels. For part segmentation, the attributes of a point in the output layer of the decoder are taken as the point features. We use the predicted object label from the classification task to guide the part segmentation task. That is, for each object class, we train a separate classifier for part segmentation of that object. Although the feature learning is unsupervised, class and segmentation labels are needed to train the classifiers for final decision-making.

Experiments We pre-trained the UFF model on the ShapeNet dataset [5] without any data augmentation. The dataset has 57,448 CAD models categorized into 55 types of man-made objects (e.g., airplane, bag, car, chair). Each CAD model is initially sampled to 2048 points, and each point has three Cartesian coordinates. We then applied the learned UFF model to other datasets without changing the filter weights (i.e., the feedforward convolutional filter weights of the Saab transform) to show the generalizability of the UFF method. This is called the pre-trained model in the following. For shape classification, we evaluated the model on the ModelNet40 dataset [37]. For part segmentation, we evaluated the model on the ShapeNetPart dataset [40], which is a subset of the ShapeNet core dataset.

Shape Classification The pre-trained model was used to obtain the shape features of the ModelNet40 dataset [37]. The point clouds were initially sampled to 2048 points. We then trained a random forest (RF) classifier, a linear SVM, and a linear least squares regressor (only the most successful results are reported here)

Table 4.5 Comparison of classification results on ModelNet40. Reproduced with permission [42].

	Method	OA (%)
Supervised[a]	PointNet [25]	89.2
	PointNet++ [26]	90.2
	PointCNN [22]	92.2
	DGCNN [35]	92.2
Unsupervised[a]	PointHop [44]	89.1
	PointHop++ [43]	91.1
Unsupervised[b]	FoldingNet [39]	88.9
	PointCapsNet [45]	88.9
	MultiTask [13]	89.1
	UFF	90.4

[a] Learning on the ModelNet40 data
[b] Transfer learning from the ShapeNet on the ModelNet40 data

on the learned features. The classification accuracy on the ModelNet40 dataset is presented in Table 4.5. The UFF model achieved an overall accuracy of 90.4%, which surpasses existing unsupervised methods. It is also competitive with state-of-the-art supervised models.

Part Segmentation We conducted experiments on the ShapeNetPart dataset [40] to predict a part category for each point. The ShapeNetPart dataset has 16,881 CAD models in total from 16 object categories, which are sampled to 2048 points to generate point clouds. Each object category is annotated with two to six parts, and there are 50 parts in total. The dataset is split into three sections: 12,137 shapes for training, 1870 shapes for validation, and 2874 shapes for testing. We followed the evaluation metric presented in [25] to calculate the mean Intersection-over-Union (mIoU) between the pointwise ground truth information and the prediction. The shape IoU was calculated by averaging the IoUs of all parts in a shape, the category mIoU was calculated by averaging over all shapes in the category, and the instance mIoU (Ins. mIoU) was computed by averaging over all shapes, while the category mIoU (Cat. mIoU) was computed by averaging over mIoUs with respect to all categories.

By following [45], 1% and 5% of the data were randomly sampled to obtain shape and point features using the pre-trained model to see both the capability of learning from limited data and the generalization ability for the segmentation task. The strategy here is to use the predicted object labels to guide the part segmentation task. Specifically, we first classified the object labels of the data using the shape feature. For each object class, we trained a random forest classifier on the extracted point features of the sampled training data. Hence, there were 16 different classifiers in total. Then, we evaluated them on the point features of the testing data with the corresponding predicted label, respectively. The part segmentation results and comparison with three state-of-the-art semi-supervised works are shown in Table 4.6. Our method had a better performance in each area.

Table 4.6 Comparison of results on the ShapeNetPart dataset with semi-supervised DNNs. Reproduced with permission [42]. Copyright © 2020, IEEE

	1% training data		5% training data	
Method[a]	OA (%)	mIoU (%)	OA (%)	mIoU (%)
SO-Net [21]	78.0	64.0	84.0	69.0
PointCapsNet [45]	85.0	67.0	86.0	70.0
MultiTask [13]	88.6	68.2	93.7	77.7
UFF	88.7	68.5	94.5	78.3

[a]Transfer learning from the ShapeNet on the ShapeNetPart data

Table 4.7 Ablation study of the object-wise segmentation. Reproduced with permission [42]. Copyright © 2020, IEEE

Object-wise	Object label	Cat. mIoU (%)	Ins. mIoU (%)
No	–	71.5	74.9
Yes	Predicted	76.2	78.3
Yes	Ground truth	78.1	81.5

To validate the object-wise segmentation method, we conducted an ablation study, the results of which are presented in Table 4.7. We randomly sampled 5% of the ShapeNetPart data following the settings above. Using the predicted test labels to guide the part segmentation task increased the Ins. mIoU by 3.4%. Besides, training multiple classifiers reduced the computation complexity.

We also compared the UFF method with other state-of-the-art supervised methods, as shown in Table 4.8. Here, we trained the model with 5% of the ShapeNetPart data (rather than using the pre-trained model). As compared with Table 4.7, pre-training does boost the performance. There is a performance gap between our model and DNN-based models. This is consistent with the classification results in Table 4.5.

Some part segmentation results of PointNet, UFF, and the ground truth labeling are visualized in Fig. 4.18. In general, the visualization results are satisfactory, although our model may fail to classify fine-grained details in some complex examples.

4.3 Registration

3D registration has been actively researched for a long time. In this book, we reviewed some of the traditional and deep learning methods for point cloud registration in Chaps. 2 and 3, respectively. The point cloud registration process can be divided into two steps—finding point correspondences and estimating the rigid transformation. The traditional methods use handcrafted local descriptors to represent the 3D points. They are derived from local surface characteristics such as the surface normal, eigenvalues, and so on. Nearest neighbor searching is used along with some additional rules to establish point correspondences and filter out bad correspondence pairs. To find the optimal transformation, methods based on

Table 4.8 Comparison of unsupervised DNNs on the ShapeNetPart dataset. The table below is the mIoU for each category correspondingly. Reproduced with permission [42]. Copyright © 2020, IEEE

Method[a]	% training data	Cat. mIoU (%)	Ins. mIoU (%)
PointNet [25]	100%	80.4	83.7
PointNet++ [26]		81.9	85.1
DGCNN [35]		82.3	85.1
PointCNN [22]		84.6	86.14
UFF	5%	73.9	76.9

Areo	Bag	Cap	Car	Chair	Ear phone	Guitar	Knife	Lamp	Laptop	Motor	Mug	Pistol	Rocket	Skate board	Table
83.4	78.7	82.5	74.9	89.6	73.0	91.5	85.9	80.8	95.3	65.2	93.0	81.2	57.9	72.8	80.6
82.4	79.0	87.7	77.3	90.8	71.8	91	85.9	83.7	95.3	71.6	94.1	81.3	58.7	76.4	82.6
84.2	83.7	84.4	77.1	90.9	78.5	91.5	87.3	82.9	96.0	67.8	93.3	82.6	59.7	75.5	82.0
84.1	86.5	86.0	80.8	90.6	79.7	92.3	88.4	85.3	96.1	77.2	95.3	84.2	64.2	80.0	83.0
71.9	68.8	74.9	68.0	84.4	78.2	86.2	76.1	67.7	94.5	58.0	93.2	67.5	49.9	68.0	75.6

[a]Learning on the ShapeNetPart data

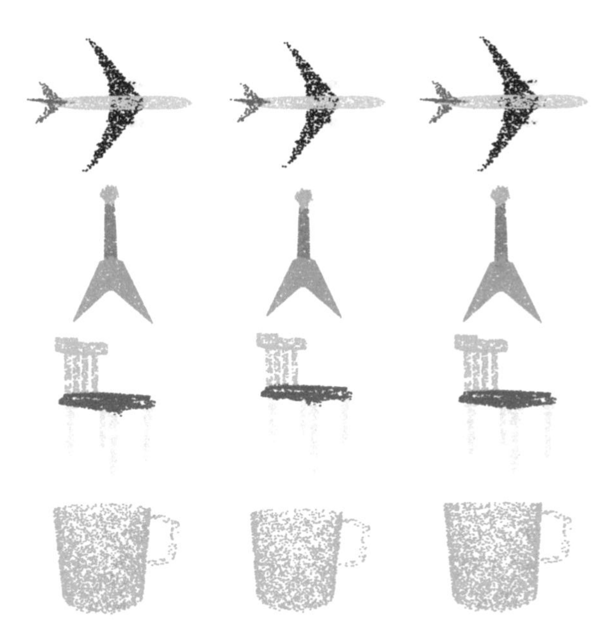

Fig. 4.18 Visualization of part segmentation results. From left to right: PointNet, UFF, ground truth. Reproduced with permission [42]. Copyright © 2020, IEEE

singular value decomposition (SVD) and least squares coupled with RANSAC are popular. In the deep-learning-based methods, the distinction between finding correspondences and transformation is sometimes blurred, because this class of methods trains end-to-end networks for predicting the transformation and/or the correspondence set.

Inspired by the successive subspace learning (SSL) design principle, particularly the PointHop and PointHop++ methods, we propose two methods for the problem of point cloud registration: Salient Points Analysis (SPA) [14] and R-PointHop [15]. These methods are discussed in detail next. SPA and R-PointHop learn 3D point features in a one-pass feedforward manner without backpropagation and offer several advantages over their traditional and deep learning counterparts.

4.3.1 Salient Points Analysis (SPA)

Salient Points Analysis (SPA) [14] extends the PointHop++ [43] classification method for the point cloud object registration task. The classifier module in PointHop++ is removed since it is not required for the registration task. This makes SPA completely unsupervised. The point features are used to find the correspondence set and estimate the 3D rotation and translation. This pipeline is very similar to the classical iterative closest point (ICP) algorithm; however, instead

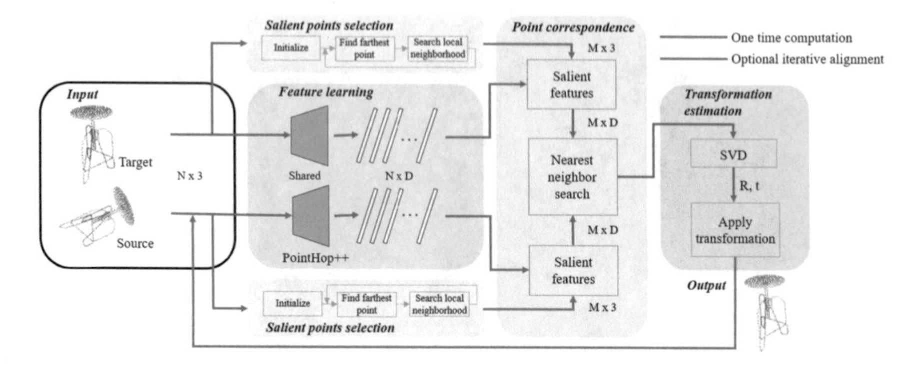

Fig. 4.19 System diagram of SPA. Reproduced with permission [14]. Copyright © 2020, IEEE

of using the point correspondences to find the closest point, SPA uses a learned local point feature to perform matching. SPA performs better than ICP, particularly when the rotation angles are larger. This will be discussed more in the experimental part. Furthermore, SPA uses the local surface properties to select a subset of salient points for finding the transformation.

The problem statement of SPA is as follows. The two point clouds to be registered are called the source and target point clouds. The source point cloud $Y \in \mathbb{R}^3$ is obtained by rotating and translating the points in the target point cloud $X \in \mathbb{R}^3$. The underlying rotation matrix is given by $R_{XY} \in SO(3)$, with translation $t_{XY} \in \mathbb{R}^3$. Given X and Y, the goal is to find the optimal rotation $R^*_{XY} \in SO(3)$ and translation $t^*_{XY} \in \mathbb{R}^3$ that minimize the pointwise mean squared error between the point correspondences (x_i, y_i) found. The error term is given by

$$E(R_{XY}, t_{XY}) = \frac{1}{N} \sum_{i=0}^{N-1} \| R^*_{XY} x_i + t^*_{XY} - y_i \|^2, \tag{4.11}$$

where N is the total number of correspondences used to find the transformation. The system diagram of SPA is shown in Fig. 4.19. There are four main modules: feature learning, salient point selection, point correspondence, and transformation estimation. We now elaborate on each of these modules.

Feature Learning The inputs to the feature learning process are the point cloud objects from the training data, consisting of N points with three coordinates. During training, the target point clouds are fed to the PointHop++ feature extractor. This is equivalent to module 1 in [43]. Although this method is similar to that of PointHop++, there is a subtle difference that promotes the choice of different model parameters: in PointHop++, the global point cloud feature is required for object-wise class prediction. However, for the registration task, a local point descriptor is desired and global structure awareness is not always useful. The point features should be similar to the features of a corresponding point in a similar local region and should be different for non-matching points. Consequently, the neighborhood size should be small while constructing attributes in every hop. Typically, a neighborhood size of 64 gives the best performance for the classification task, while in SPA, we restrict the neighborhood size to 8 in each hop to preserve locality.

Similar to PointHop++, four hops are considered, and all leaf nodes are collected as point features. The feature dimension D depends on the energy threshold parameter set for PointHop++. In contrast to PointHop++, the point cloud downsampling operation between two hops is avoided, and features are found for all points. The same intuition is used here. Point cloud downsampling using farthest point sampling (FPS) was adopted in PointHop and PointHop++ for growing the receptive field and capturing the global structure information. As this contradicts the motive of the registration task, downsampling is not performed in SPA. Following this, the feature aggregation step used in PointHop++ is also not included. While the aggregation operation gives the global point cloud feature vector, we are only interested in the per-point features here. Similarly, the classifier is omitted since it is not required here.

These modifications make the SPA method fully unsupervised. Similar to the classification case, the features are learned in a feedforward and one-pass manner. After training, this module takes in a point cloud with N points and outputs a D-dimensional feature vector for every point. During registration, both the source and target point clouds are fed to the same trained model. The feature learning process is shaded blue in Fig. 4.19.

Salient Point Selection The salient point selection process uses the local geometric properties of points to find a subset of discriminant points. Later, these points and their features are used to find the 3D rigid transformation. A fixed number M of salient points are found for both the target and source point clouds. The procedure is as follows. Principal components analysis (PCA) in a local region is a good indicator of local surface characteristics. For every point p_i, its K nearest neighbors are found as $p_{i1}, p_{i2}, \cdots, p_{iK}$. The inputs to the local PCA operation are the 3D coordinates of these K points, given by a $K \times 3$ matrix of local coordinates. The 3×3 covariance matrix is then calculated, whose eigen decomposition yields three non-negative eigenvalues. The smallest eigenvalue is stored for every point. Points lying on smooth or flat surfaces have a smaller third eigenvalue, because they can be represented using two principal components in the transformed domain without much loss of information. The salient points are those having a higher energy for the third principal component (or the third eigenvalue being relatively larger). Accordingly, points in the target and source point clouds are sorted by the third eigenvalue in descending order. However, selecting the top M points from this ordered list may lead to points being selected from only a few specific regions of the point cloud. Instead, spatially disperse points are desired, along with salient behavior. To achieve both goals, the eigenvalue information is coupled with FPS, and each salient point is iteratively selected from the farthest region from the current salient point. The first salient point is initialized to be the point with the largest third

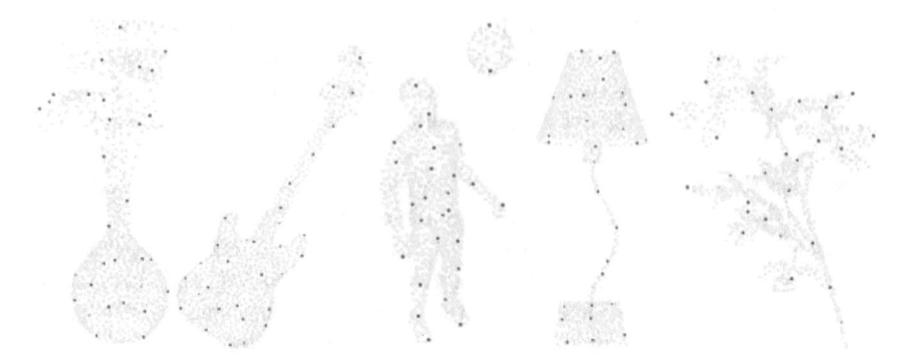

Fig. 4.20 Salient points selected by SPA for several point clouds are highlighted. Reproduced with permission [14]. Copyright © 2020, IEEE

eigenvalue.

$$m_0 = p_i, \quad s.t.\ i = \arg\max_j(\lambda_j). \tag{4.12}$$

To find the next salient point m_1, the farthest point of m_0 is found. The points in the local neighborhood of this point are then analyzed. The point with the largest third eigenvalue among these neighboring points is selected to be the next salient point. This process is repeated until M salient points are found. Salient points for some point clouds selected by SPA are highlighted in red in Fig. 4.20.

Point Correspondence The subset of M salient points is found from the target and source point clouds. The respective features of these points are obtained using the feature extraction module. Next, we find the set of correspondences by nearest neighbor searching. For every salient point in the target point cloud, its features are compared with the features of all the source point clouds. The point in the source cloud with the smallest l_2 distance in D-dimensional feature space is selected to be the matching point. The ordered pair of corresponding points is given by (x_i, y_i) and there are M such pairs.

Transformation Estimation The final step in the registration process is to use the correspondences to predict the transformation that best aligns the source with the target. It is found using singular value decomposition (SVD) as follows. First, the centroids of the two point clouds are found from the correspondence set as

$$\bar{x} = \frac{1}{N}\sum_{i=0}^{N} x_i, \quad \bar{y} = \frac{1}{N}\sum_{i=0}^{N} y_i. \tag{4.13}$$

Then, the covariance matrix is calculated as

$$Cov(X, Y) = \sum_{i=0}^{N}(x - x_i)(y - y_i)^T. \tag{4.14}$$

SVD of the covariance matrix gives $Cov(X, Y) = USV^T$, where U is the matrix of left singular vectors, V is the matrix of right singular vectors, and S is the diagonal matrix of singular values. The optimal rotation matrix and the translation vector are then given by

$$R^*_{XY} = VU^T, \quad t^*_{XY} = -R^*_{XY}\bar{x} + \bar{y}. \tag{4.15}$$

Iterative Alignment The source is aligned to the target point cloud using the transformation obtained from Eq. 4.15. Optionally, an iterative alignment (similar to ICP) can be performed, wherein the updated source point cloud acts as the new source for the subsequent iteration, and the process continues. In this case, the target point cloud is fixed, and its features and salient points are found only once. Even when performing such an iterative alignment, the PointHop++ model is trained only once. The iterative portion of the method is marked by the blue lines in Fig. 4.19. Some successful registration results using SPA are depicted in Fig. 4.21.

Experiments Point cloud objects from the ModelNet40 dataset were used to train and evaluate the SPA method. The choice of hyperparameters for training the feature extractor (PointHop++) is as follows. The number of nearest neighbors considered in the first hop was set to 32, while 8 nearest neighbors were used in the subsequent hops. The energy threshold parameter T was kept as 0.0001. The evaluation metrics used were mean squared error (MSE), root mean squared error (RMSE), and mean

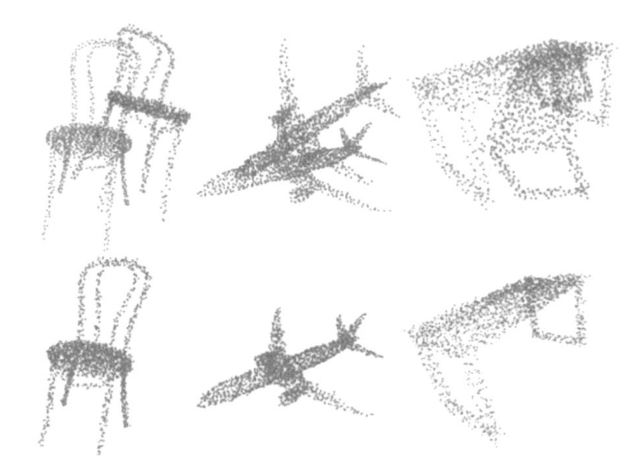

Fig. 4.21 Registration of noisy point clouds (in orange) with noiseless point clouds (in blue) using SPA. Reproduced with permission [14]. Copyright © 2020, IEEE

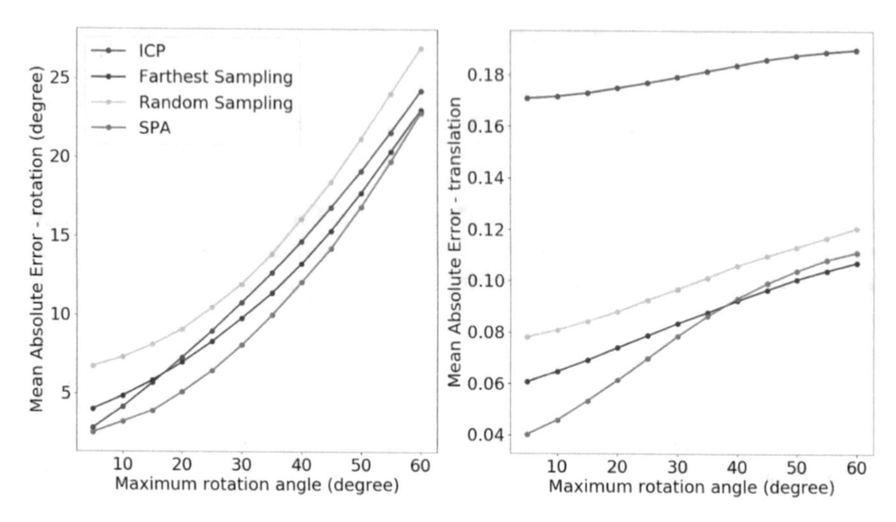

Fig. 4.22 Comparison of the mean absolute registration errors of rotation and translation as a function of the maximum rotation angle. Reproduced with permission [14]. Copyright © 2020, IEEE

absolute error (MAE) for the rotation angles (along three coordinate axes) and the translation vector. In the first experiment, the entire training dataset was used during training. During testing (registration), a random rotation and translation were applied to the target point cloud to generate the source point cloud. The rotation angles along the X, Y, and Z axes were uniformly sampled between 0° and a maximum angle, which was varied from 5° to 60° in 5° intervals. Translation along the three axes was uniformly sampled in $[-0.5, 0.5]$.

The MAE for rotation and translation is plotted as a function of maximum rotation angle in Fig. 4.22. SPA consistently performs better than ICP in terms of rotation and translation errors. The graph also shows two more cases where the salient point selection rule is replaced with random point sampling and farthest point sampling. Both cases perform worse than the salient point method, thereby justifying its use.

The next set of experiments is on unseen classes and noisy point clouds. The experimental setup is as follows. For the unseen class case, the SPA method was trained on target point clouds from the first 20 classes of ModelNet40. During registration, the objects belonging to other 20 classes were used. This experiment tests the generalizability of SPA toward untrained classes. For the noisy point cloud case, all 40 classes were used during training and testing. A Gaussian noise of zero mean and 0.01 standard deviation was added to the source. This experiment explores the noise resilience of SPA. For both experiments, the three rotation angles were uniformly sampled in $[0, 45^\circ]$, and the translation along each axis was uniform in $[-0.5, 0.5]$.

The results of the SPA method were compared with those of ICP and deep-learning-based methods PointNetLK [2] and DCP [33]. The results are presented in Table 4.9. SPA performs better than ICP and is comparable with PointNetLK; however, it is inferior to DCP. In addition, the performance does not drop sharply for the noisy case, indicating SPA is robust to noise.

Error Analysis A histogram of the mean absolute rotation error for the experiment on unseen classes is shown in Fig. 4.23. The class-wise error distribution for the 20 classes is also shown. From the histogram, we can infer that a large number of point clouds align well with very small error. In particular, 38% of the point clouds have a mean absolute rotation error of less than 1° and 72% of objects have an error of less than 5°. In addition, 8 of the 20 classes have an average error of less than 5°. This indicates that there are fewer corner cases that are contributing to a large error.

Some shortcomings of SPA were identified and addressed in the follow-up work R-PointHop. Nevertheless, SPA still offers a promising direction for point cloud registration using the SSL approach. The model size of SPA is only 64 kB, which is far smaller than that of DCP (21 MB). In addition, the training time is merely 30 min without GPU usage. Deep learning methods offer superior performance over traditional methods, but at the cost of model size and training times. In resource-constrained situations, SPA can offer a good trade-off between registration performance and resource utilization.

4.3.2 R-PointHop

SPA demonstrated the usefulness of features learned using the SSL methodology for point cloud registration tasks. However, the performance was subpar compared to that of supervised learning methods. The follow-up work, R-PointHop [15], was designed with consideration to the issues that hamper the performance of SPA. Initially, it was observed that SPA fails when the rotation angles are large. This issue closely resembles the case where ICP does not align well without proper initialization. Although SPA uses near-to-far neighborhood information to construct features instead of pure 3D coordinate information like in ICP, it still fails beyond a certain rotation angle. This shows the necessity of a global alignment method. R-PointHop addresses this issue by incorporating local reference frames for points, thereby allowing rotation-invariant feature learning.

Furthermore, SPA makes an implicit assumption that the complete source and target point clouds are available. However, this is seldom true. In most practical cases, only partial views of the source and target are available, and only a subset of points from the two point clouds overlaps. SPA cannot handle cases where there is a partial view, thereby limiting its capability. R-PointHop achieves partial registration by replacing the geometry-based salient point selection process with a feature-based selection process that uses point feature information to identify overlapping points. Indeed, detailed experiments have highlighted the effectiveness of R-PointHop for 3D registration tasks.

Table 4.9 Registration performance comparison on ModelNet-40 with respect to unseen classes (left) and noisy input point clouds (right). Reproduced with permission [14]. Copyright © 2020, IEEE

	Registration errors on unseen classes						Registration errors on noisy input point clouds					
Method	MSE (R)	RMSE (R)	MAE (R)	MSE (t)	RMSE (t)	MAE (t)	MSE (R)	RMSE (R)	MAE (R)	MSE (t)	RMSE (t)	MAE (t)
ICP [3]	467.37	21.62	17.87	0.049722	0.222831	0.186243	558.38	23.63	19.12	0.058166	0.241178	0.206283
PointNetLK [2]	306.32	17.50	5.28	0.000784	0.028007	0.007203	256.16	16.00	4.60	0.000465	0.021558	0.005652
DCP [33]	19.20	4.38	2.68	0.000025	0.004950	0.003597	6.93	2.63	1.52	0.000003	0.001801	0.001697
SPA	354.57	18.83	6.97	0.000026	0.005120	0.004211	331.73	18.21	6.28	0.000462	0.021511	0.004100

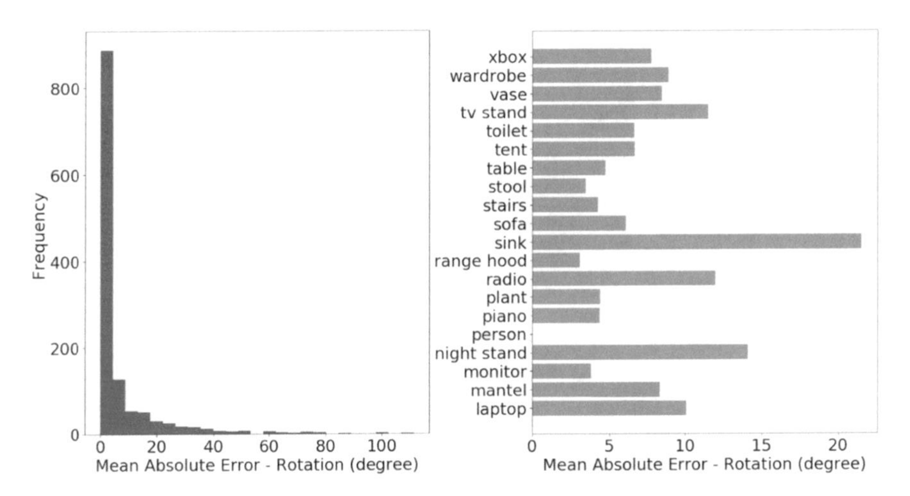

Fig. 4.23 Error histogram (left) and class-wise error (right). Reproduced with permission [14]. Copyright © 2020, IEEE

The problem statement of R-PointHop is similar to that of SPA, with a relaxation of the source and target to facilitate partial region of the 3D object. The goal is to find a rigid transformation comprising 3D rotation and translation that optimally aligns the source $G \in \mathbb{R}^3$ and target, $F \in \mathbb{R}^3$. The source is obtained by rotating the target by $R \in SO(3)$ and translating it by $t \in \mathbb{R}^3$. Here, $SO(3)$ is the special orthogonal 3D rotation group in Euclidean space. Given F and G, the optimal rotation $R^* \in SO(3)$ and the translation $t^* \in \mathbb{R}^3$ to minimize the mean squared error E between the matching points are found, where

$$E(R, t) = \frac{1}{N} \sum_{i=0}^{N-1} \|R^* f_i + t^* - g_i\|^2. \tag{4.16}$$

Here, (f_i, g_i) is the selected pair of N corresponding points.

Local Reference Frame Principal component analysis (PCA) of points in a local region provides information about the local surface structure. It does not depend on the absolute 3D coordinates of points and is invariant under rotation and translation. Hence, it is expected that matching points and surfaces should have similar local PCA. A local PCA computation is used in SPA to select a subset of salient points; however, SPA considers the eigenvalue of the point instead of the eigenvector.

The local reference frame (LRF) of a point is derived as follows. The K nearest neighbors of a point are found and PCA is conducted on their 3D coordinates. This gives three mutually orthogonal eigenvectors. The eigenvectors are sorted in descending order of the associated eigenvalues. X, Y, and Z are used as a convention to represent the original coordinate system in which the points are defined. For the LRF, the three axes corresponding to the three eigenvectors with the largest, middle, and smallest eigenvalues are labeled P, Q, and R, respectively. Each

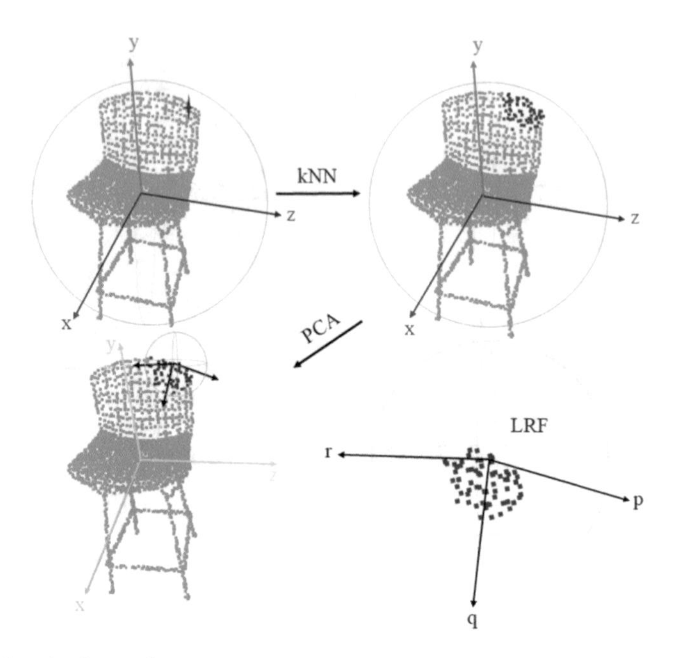

Fig. 4.24 Local reference frame (LRF) [15]. Permitted by CC BY 4.0 License

eigenvector is associated with a sign ambiguity, because the negative vector is also a valid eigenvector. The distribution of neighborhood points at every hop is used to resolve this ambiguity. This is discussed further later. Then, the positive eigenvectors for each point are represented as (p^+, q^+, r^+) and the negative eigenvectors as (p^-, q^-, r^-). These directions are unique for every point. An example LRF is shown in Fig. 4.24.

Point Attribute Construction For every point in a point cloud, a local attribute is constructed as follows. First, the K nearest neighbors of the point are found. Depending on the size of the point cloud, these may be the same as those for the LRF, or a larger neighborhood may be considered. For every point, the XYZ coordinates of the neighboring points are projected on the LRF of the current point. The eigenvectors (p^+, q^+, r^+) are considered as the default axes.

The sign ambiguity is resolved separately for every eigenvector as follows. First, the 1D coordinates of K points along every eigenvector are considered. Next, the median point is found, and the first-order moments about the median point are calculated for the points to the left and right of the median. The left and right moments are given by

$$M_p^l = \sum_i |p_i - p_m| \quad \forall\, p_i < p_m \tag{4.17}$$

$$M_p^r = \sum_i |p_i - p_m| \quad \forall\, p_i > p_m, \tag{4.18}$$

where p_i is the 1D coordinates of point i projected to p^+ and p_m is the projected value of the median point. If $M_p^r > M_p^l$, the original assignment is retained as p^+, whereas if $M_p^r \leq M_p^l$, the assignment is swapped to p^- to ensure that the direction with a larger first-order moment is along the positive axis. The same procedure is repeated for axes Q and R. This operation is implemented as post-multiplying the local data matrix of dimension $K \times 3$ with a diagonal reflection matrix, $R' \in R^{3\times3}$, whose diagonal elements are either 1 or -1 depending on the chosen sign. The matrix is defined as

$$R'_{ii} = \begin{cases} 1, & \text{if } M_i^l < M_i^r, \\ -1, & \text{otherwise}, \end{cases} \tag{4.19}$$

and

$$R'_{ij} = 0, \quad \text{if } i \neq j. \tag{4.20}$$

The 3D space of the K nearest neighbor points is divided into eight octants using the local reference frame centered at the current point. For each octant, the mean of the 3D coordinates of points in that octant is calculated, and the eight means are concatenated to form a 24-dimensional (24D) vector. This is taken as the attribute of the current point. The octant division and grouping are similar to that of PointHop; however, PointHop uses the XYZ axes as reference at all times, whereas R-PointHop employs a local reference frame. The attribute construction process is shown in Fig. 4.25.

Multi-hop Features The Saab transform [43] is conducted on all the 24D point attributes from the training data to obtain their 24D spectral representations. This is the output of the first hop. The energy of each node is calculated as in [43], and the nodes with an energy greater than the threshold energy T are passed to the next hop. The rest of the nodes are discarded. Since the first hop features carry more local structure information, which may be similar in different regions of the point cloud, these nodes may lead to mismatched correspondences. Hence, these nodes are discarded, unlike in the original PointHop++ methodology.

The point cloud is then downsampled using farthest point sampling (FPS) before moving on to the second hop. FPS ensures that the structure of the point cloud is

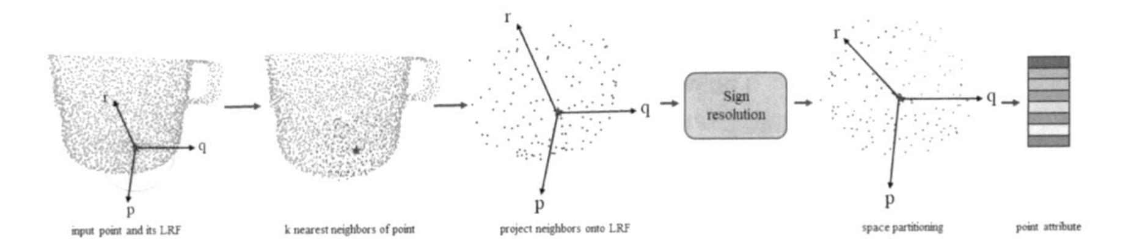

Fig. 4.25 Point attribute construction [15]. Permitted by CC BY 4.0 License

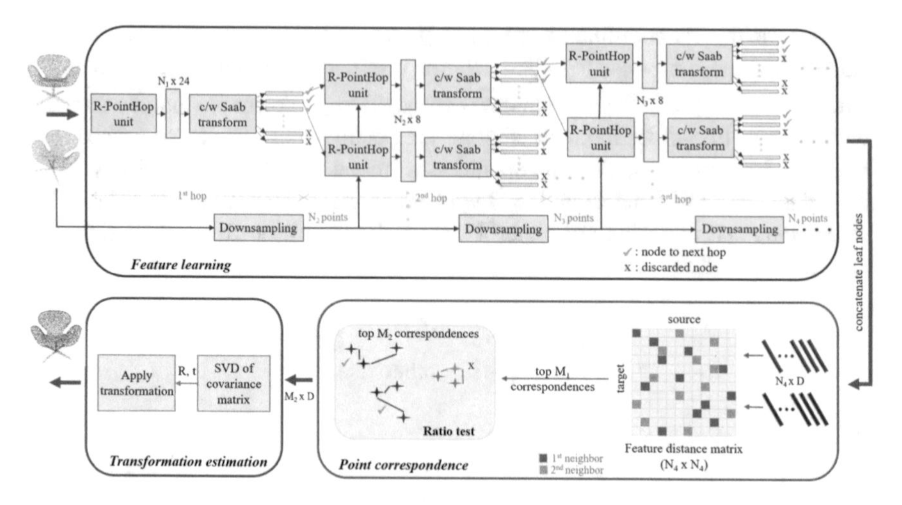

Fig. 4.26 System diagram of R-PointHop [15]. Permitted by CC BY 4.0 License

preserved after downsampling. It also helps to reduce computations and grow the receptive field quickly. In the second hop, the attribute construction process is the same for the nodes passed from the first hop. Since the spectral components are uncorrelated, they are processed separately using the channel-wise Saab transform [43] for the second and subsequent hops. The K nearest neighbors of the current point are found. Note that these are different from the first hop neighbors due to point cloud downsampling. The 3D coordinates of the neighbors are then projected onto the local reference frame of the current point. Since the set of nearest neighbors is different from that in the first hop, the eigenvector direction is recalculated. Points are divided into eight octants based on the local reference frame. The mean of the 1D features of all points in each octant is calculated, and then the eight means are concatenated to obtain 8D second hop attributes for a node. All point attributes are collected for each node, and the channel-wise Saab transform is used to obtain an 8D spectral representation. This process is repeated for all nodes in the second hop.

The multi-hop feature learning process is continued for four hops. All 1D spectral components at the end of the fourth hop are concatenated to obtain the feature vectors of each point. The final feature dimension depends on the selected parameters, such as the neighborhood size, the number of points after every downsampling operation, and energy threshold for the channel-wise Saab transform. The architecture of the feature learning process is shown in Fig. 4.26.

Point Correspondences The point feature extraction process is completely agnostic of the registration process. Once the R-PointHop model is trained for registration, the target and source points are fed to the same model to extract pointwise features. The feature distance matrix is then constructed. If both the source and target point clouds comprise N points, then the distance matrix is of size $N \times N$, whose ij-th element is the l_2 distance between i-th point in the target point cloud and j-th

Fig. 4.27 Correspondences found using R-PointHop [15]. Permitted by CC BY 4.0 License

point in the source point cloud. The closest matching point in the source point cloud for every point in the target point cloud is found by obtaining the minimum along every row. These pairs of points are used as an initial set of correspondences. Furthermore, these correspondences are ranked based on the l_2 distance between the features of matching points. The top M_1 matches have the smallest l_2 distances. This step is crucial for partial registration. Suppose only a subset of points between two partial point clouds overlaps. In such a case, all the non-overlapping points would potentially have a larger feature l_2 distance with their matches. To further refine this subset of M_1 correspondences, a ratio test is performed. Here, the ratio between the distance to the first neighbor in the feature space and the distance to the second neighbor is calculated. For a perfect match, the feature distance to the first neighbor is very close to zero. Hence, point correspondences with a lower ratio are generally more reliable. The correspondences are then ranked based on this ratio, and the top M_2 correspondences are selected. Therefore, unlike SPA, which uses local geometry information to select salient points, R-PointHop utilizes rich feature information to select a high-quality subset of correspondences.

The point correspondence step is shown in the R-PointHop block diagram in Fig. 4.26. Some of the correspondences found using R-PointHop are marked in Fig. 4.27.

Transformation Estimation The correspondences $(\mathbf{f}_i, \mathbf{g}_i)$ found in the previous step are used to estimate the optimal rotation R^* and translation t^* that minimize the error function discussed in Eq. 4.16. This is solved numerically using singular value decomposition (SVD) of the data covariance matrix, as previously discussed for SPA (Sect. 4.3.1). The procedure is summarized again below.

1. The centroids are found from the correspondences as

$$\bar{\mathbf{f}} = \frac{1}{N}\sum_{i=0}^{N-1} \mathbf{f}_i, \quad \bar{\mathbf{g}} = \frac{1}{N}\sum_{i=0}^{N-1} \mathbf{g}_i. \tag{4.21}$$

2. Then, the covariance matrix is computed as

$$Cov(F, G) = \sum_{i=0}^{N-1} (\mathbf{f}_i - \bar{\mathbf{f}})(\mathbf{g}_i - \bar{\mathbf{g}})^T. \tag{4.22}$$

3. The covariance matrix is decomposed into U, S, and V:

$$Cov(F, G) = USV^T, \tag{4.23}$$

   where U is the matrix of left singular vectors, S is the diagonal matrix containing singular values, and V is the matrix of right singular vectors. In the current case, $U,\ S$, and V are 3×3 matrices.
4. Then, the optimal rotation matrix R^* is given by

$$R^* = VU^T, \tag{4.24}$$

   and the optimal translation vector t^* is found using R^* and the centroids $\bar{x}$ and $\bar{y}$ as

$$t^* = -R^*\bar{\mathbf{f}} + \bar{\mathbf{g}}. \tag{4.25}$$

Finally, the aligned source point cloud (G') is found using R^* and t^* as

$$G' = R^{*T}(G - t^*), \tag{4.26}$$

where R^{*T} is the transpose of R^* that applies the inverse transformation. Unlike SPA, R-PointHop is non-iterative. Some successful registration results using R-PointHop are pictured in Fig. 4.28.

Experiments The ModelNet40 dataset was used for training and evaluations. R-PointHop was compared with six methods: ICP [3], Go-ICP [38], FGR [46], PointNetLK [2] DCP [33], and SPA. A random rotation was applied to the target

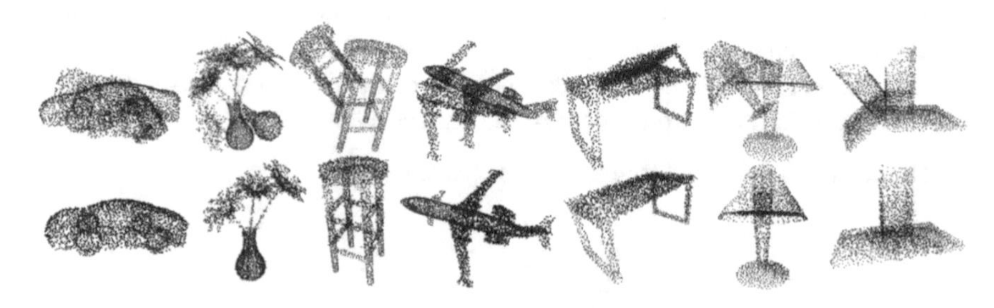

Fig. 4.28 Registration of point clouds from the ModelNet40 dataset using R-PointHop [15]. Permitted by CC BY 4.0 License

point cloud about the XYZ coordinate axes. The rotation angles along each axis are uniformly sampled in $[0°, 45°]$. Then, a random translation was applied along the three axes, which are uniform in $[-0.5, 0.5]$, to obtain the source point cloud. During training, just the target point clouds are used. For evaluation, the mean square error (MSE), root mean square error (RMSE), and mean absolute error (MAE) between the ground truth information and predicted rotation angles translation vector are reported. Furthermore, experiments on real-world point clouds from the Stanford 3D scanning repository [11, 17, 31] were performed. The latter experiments show that R-PointHop can be used as an initialization for ICP, as it shows some useful properties of 3D point descriptors.

Registration on Unseen Data For this experiment, R-PointHop was trained on training samples of all 40 classes of the ModelNet40 dataset. During registration, point clouds from the test data were used. The results are presented in Table 4.10. Notably, R-PointHop outperformed all six benchmarking methods.

Registration on Unseen Classes In this experiment, only the first 20 classes of the ModelNet40 dataset were used during training. During registration, test samples from the remaining 20 classes were used. Referring to the results in Table 4.11, it can be seen that R-PointHop can generalize well on unseen classes. In comparison, PointNetLK and DCP have relatively larger errors compared to their errors in Table 4.10. This implies that these methods are biased to categories that have been seen previously.

Registration on Noisy Point Clouds In this experiment, a noisy source point cloud was aligned with a target. To produce noisy source point clouds, a Gaussian noise with zero mean and a standard deviation of 0.01 was added. The registration results are tabulated in Table 4.12. The results show that R-PointHop is robust to Gaussian noise. A fine alignment step using ICP further reduces the error, suggesting that R-PointHop can achieve coarse alignments in the presence of noise.

Registration on Real-World Data R-PointHop was tested on 3D point clouds from the Stanford Bunny dataset [31]. A subset of 2048 points was randomly sampled from the bunny models, such that the selected points were evenly spanned

Table 4.10 Registration on unseen point clouds [15]. Permitted by CC BY 4.0 License

Method	MSE (R)	RMSE (R)	MAE (R)	MSE (t)	RMSE (t)	MAE (t)
ICP [3]	451.11	21.24	17.69	0.049701	0.222937	0.184111
Go-ICP [38]	140.47	11.85	2.59	0.00659	0.025665	0.007092
FGR [46]	87.66	9.36	1.99	0.000194	0.013939	0.002839
PointNetLK [2]	227.87	15.09	4.23	0.000487	0.022065	0.005405
DCP [33]	1.31	1.14	0.77	0.000003	0.001786	0.001195
SPA [14]	318.41	17.84	5.43	0.000022	0.004690	0.003261
R-PointHop	0.12	0.34	0.24	0.000000	0.000374	0.000295

Table 4.11 Registration on unseen classes [15]. Permitted by CC BY 4.0 License

Method	MSE (R)	RMSE (R)	MAE (R)	MSE (t)	RMSE (t)	MAE (t)
ICP [3]	467.37	21.62	17.87	0.049722	0.222831	0.186243
Go-ICP [38]	192.25	13.86	2.91	0.000491	0.022154	0.006219
FGR [46]	97.00	9.84	1.44	0.000182	0.013503	0.002231
PointNetLK [2]	306.32	17.50	5.28	0.000784	0.028007	0.007203
DCP [33]	9.92	3.15	2.01	0.000025	0.005039	0.003703
SPA [14]	354.57	18.83	6.97	0.000026	0.005120	0.004211
R-PointHop	0.12	0.34	0.25	0.000000	0.000387	0.000298

Table 4.12 Registration on noisy point clouds [15]. Permitted by CC BY 4.0 License

Method	MSE (R)	RMSE (R)	MAE (R)	MSE (t)	RMSE (t)	MAE (t)
ICP [3]	558.38	23.63	19.12	0.058166	0.241178	0.206283
Go-ICP [38]	131.18	11.45	2.53	0.000531	0.023051	0.004192
FGR [46]	607.69	24.65	10.05	0.011876	0.108977	0.027393
PointNetLK [2]	256.15	16.00	4.59	0.000465	0.021558	0.005652
DCP [33]	1.17	1.08	0.74	0.000002	0.001500	0.001053
SPA [14]	331.73	18.21	6.28	0.000462	0.021511	0.004100
R-PointHop	7.73	2.78	0.98	0.000001	0.000874	0.003748
R-PointHop + ICP	1.16	1.08	0.21	0.000001	0.000744	0.001002

Table 4.13 Registration on the Stanford Bunny dataset [15]. Permitted by CC BY 4.0 License

Method	MSE (R)	RMSE (R)	MAE (R)	MSE (t)	RMSE (t)	MAE (t)
ICP [3]	177.35	13.32	10.72	0.0024	0.0492	0.0242
Go-ICP [38]	166.85	12.92	4.52	0.0018	0.0429	0.0282
FGR [46]	3.98	1.99	1.49	0.0397	0.1993	0.1658
DCP [33]	41.45	6.44	4.78	0.0016	0.0406	0.0374
R-PointHop	2.21	1.49	1.09	0.0039	0.0361	0.0269

across the object. The R-PointHop model derived from all 40 classes of the ModelNet40 dataset was used during feature extraction. The results are compared with those of other methods in Table 4.13. One registration result is visualized in Fig. 4.29. From Table 4.13, we can infer that R-PointHop trained on the ModelNet40 dataset performs well on the Stanford Bunny dataset. Furthermore, experimental results on point clouds from the Stanford 3D scanning repository, which includes Bunny, Buddha [11], Dragon [11], Armadillo [17], and so on, are shown in Fig. 4.30.

Registration on Partial Data Partial-to-partial registration experiments were performed on source and target point clouds having only a subset of points in common. The number of overlapping points between the source and target point clouds was between 512 and 768. The results of the partial-to-partial registration

Fig. 4.29 Registration on the Stanford Bunny dataset [15]. Permitted by CC BY 4.0 License

Fig. 4.30 Registration of point clouds from the Stanford 3D scanning repository [15]. Permitted by CC BY 4.0 License

tests are presented in Table 4.14. Two scenarios are considered, similarly to the previous experiments, that is, registration on unseen point clouds and registration on unseen classes. R-PointHop works well for the partial case. By using the feature l_2 distance and the ratio test to select the top correspondences, matching of the overlapping points is enhanced while ignoring non-overlapping points. The row of R-PointHop* in Table 4.14 shows the error when the ratio test is not performed. The error is reduced when including the ratio test, thereby justifying its inclusion.

Local vs. Global Registration ICP is a local registration algorithm, which means that it only works well when the optimal alignment is close to the initial alignment. R-PointHop can be used as an initialization for ICP, whereby it can achieve an adequate initial global alignment. Then, ICP can be used to fine-tune the alignment. The MAE and RMSE for rotation and translation are plotted against the maximum rotation angle in Fig. 4.31. As the maximum rotation angle in increased, the MAE and RMSE for ICP increase steadily. However, the RMSE and MAE for R-PointHop are almost constant. This shows the global registration capability of R-PointHop.

Table 4.14 Registration on partial point clouds (R-PointHop* indicates choosing correspondences without the ratio test) [15]. Permitted by CC BY 4.0 License

	Registration errors on unseen objects						Registration errors on unseen classes					
Method (R)	MSE (R)	RMSE (R)	MAE (t)	MSE (t)	RMSE (t)	MAE (R)	MSE (R)	RMSE (R)	MAE (t)	MSE (t)	RMSE (t)	MAE
ICP [3]	1134.55	33.68	25.05	0.0856	0.2930	0.2500	1217.62	34.89	25.46	0.0860	0.293	0.251
Go-ICP [38]	195.99	13.99	3.17	0.0011	0.0330	0.0120	157.07	12.53	2.94	0.0009	0.031	0.010
FGR [46]	126.29	11.24	2.83	0.0009	0.0300	0.0080	98.64	9.93	1.95	0.0014	0.038	0.007
PointNetLK [2]	280.04	16.74	7.55	0.0020	0.0450	0.0250	526.40	22.94	9.66	0.0037	0.061	0.033
DCP [33]	45.01	6.71	4.45	0.0007	0.0270	0.0200	95.43	9.77	6.95	0.0010	0.034	0.025
PR-Net [34]	10.24	3.12	1.45	0.0003	0.0160	0.0100	15.62	3.95	1.71	0.0003	0.017	0.011
R-PointHop*	3.58	1.89	0.11	0.0002	0.0150	0.0008	3.75	1.94	0.12	0.0002	0.0151	0.0008
R-PointHop	2.75	1.66	0.09	0.0002	0.0149	0.0008	2.53	1.59	0.08	0.0002	0.0148	0.0008

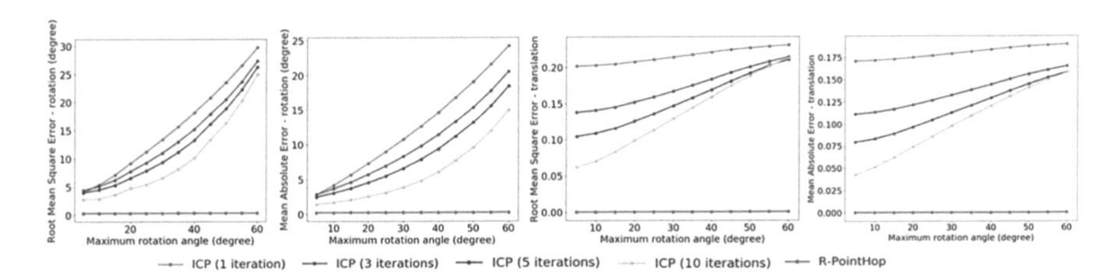

Fig. 4.31 Plots of maximum rotation angle vs. root mean square rotation error, mean absolute rotation error, root-mean-square translation error, and mean absolute translation error [15]. Permitted by CC BY 4.0 License

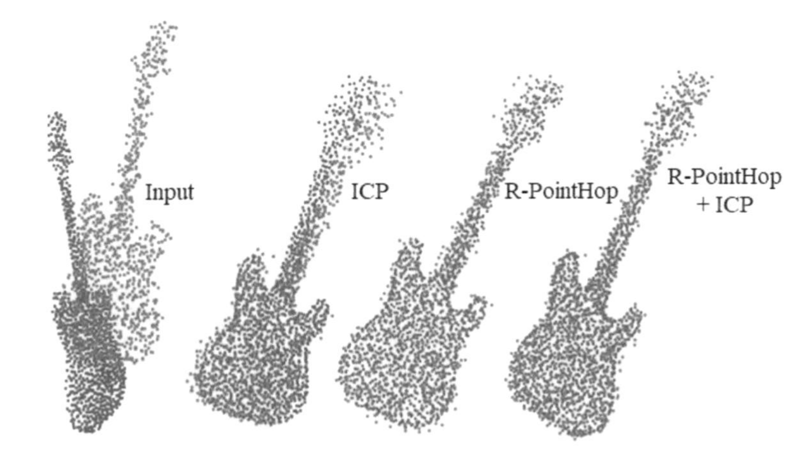

Fig. 4.32 (From left to right) The source and target point clouds to be aligned, and registration with ICP only, R-PointHop only, and R-PointHop followed by ICP [15]. Permitted by CC BY 4.0 License

In Fig. 4.32, the registration results using ICP alone, R-PointHop alone, and R-PointHop followed by ICP are shown. The best results are obtained in the third case. The ICP-alone case had the worst results, which was to be expected due to the large initial rotation.

3D Descriptor A t-distributed stochastic neighbor embedding (t-SNE) plot of several point features found by R-PointHop is shown in Fig. 4.33. The points with similar local structures are clustered together in the t-SNE plot, irrespective of their spatial locations in a 3D point cloud model or the object class. For instance, point cloud models of a table and a chair are shown on the left. The points on their legs have a similar local neighborhood and, accordingly, their features are closer in the t-SNE embedding space. The same is true of other objects in Fig. 4.33. This observation highlights the possibility of R-PointHop functioning as a general 3D descriptor. Furthermore, registrations of different objects of two object classes (airplanes and cars) are shown in Fig. 4.34. There are two different airplanes and cars. Although the objects are different, R-PointHop is still able to align the two

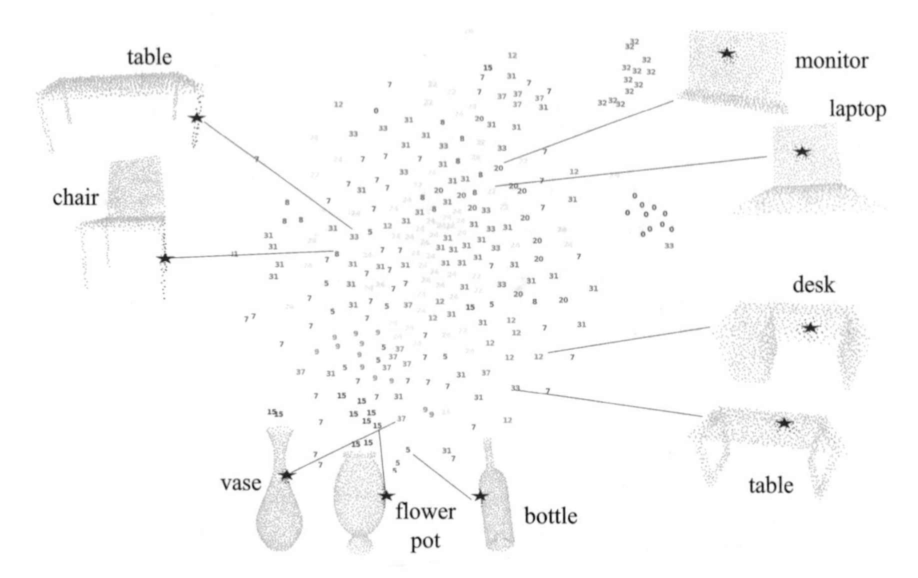

Fig. 4.33 t-SNE plot of point features [15]. Permitted by CC BY 4.0 License

Fig. 4.34 Registration of different point cloud models [15]. Permitted by CC BY 4.0 License

airplanes and two cars reasonably well. This reinforces the claim of it being a good 3D descriptor, since correspondences are found in similar parts of the airplane and car.

Green Learning Deep learning methods tend to have a large model size, thereby making them difficult to deploy on mobile devices. The training of deep learning methods has a large carbon footprint. Alongside the environmental impact, expensive computational resources such as GPUs are needed to train deep networks in reasonable time. The research community is in search of environmentally friendly solutions to different AI tasks, and Green AI [30] is on the rise.

R-PointHop is a green solution due to its small model size and reduced training time compared with deep-learning-based methods. Some observations on training times are as follows:

- DCP takes about 27.7 h to train using eight NVIDIA Quadro M6000 GPUs.
- PointNetLK takes approximately 40 min to train one epoch using one GPU, while the default training setting is 200 epochs. Thus, the total training time can be as high as 133.33 h.
- Instead, R-PointHop only takes 40 min to train all model parameters using one Intel(R) Xeon(R) CPU E5-2620 v3 at 2.40 GHz.

The model size of R-PointHop is only 200 kB, while that of PointNetLK and DCP is 630 kB and 21.3 MB, respectively. R-PointHop offers impressive benefits when these factors are considered along with the registration performance.

4.4 Other Applications of Successive Subspace Learning

Aside from point cloud processing, successive subspace learning (SSL) principles have been applied to several other 2D vision tasks, such as facial classification, Deepfake detection, anomaly detection, texture synthesis, image super-resolution, image forensics, medical image processing, and so on. The PixelHop and PixelHop++ methods form the backbone of these works. In this section, we discuss three such methods, namely FaceHop, DefakeHop, and AnomalyHop, to give a glimpse of the research using SSL.

4.4.1 FaceHop

FaceHop [29] is a lightweight method for gender classification from low -resolution images of faces. It is designed to meet the requirements of edge/mobile computing systems and other resource-constrained environments such as rescue missions and field operations in remote locations. FaceHop follows the traditional pattern recognition paradigm and decouples the process of feature extraction and decision. For feature extraction, the PixelHop++ method is used instead of any handcrafted features. The gender classification pipeline of FaceHop is illustrated in Fig. 4.35.

As shown in Fig. 4.35, FaceHop consists of four modules: preprocessing, PixelHop++, feature extraction, and classification. The image preprocessing operations include face landmark detection, image rotation to reduce the effect of pose variation, centering and cropping to reduce background, histogram equalization,

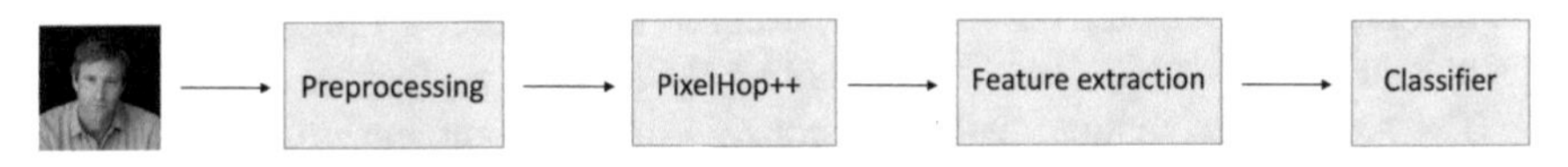

Fig. 4.35 Block diagram of FaceHop. Reproduced with permission [28]. Copyright © 2021, Springer Nature Switzerland AG

and finally, resizing the image to a size of 32×32. These images are then fed to PixelHop++ to extract the hierarchical features that represent the short-, mid-, and long-range neighborhoods of the pixels. Three hops are conducted. Each hop provides complementary information. The first hop has a response map of size 28×28 and provides a spatially detailed representation of the input. The second hop with a response map of size 10×10 gives a much coarser view and does not include face details like the first hop. The final hop loses all spatial details and provides a single value at a frequency channel that covers the entire face. The popular eigenface approach captures only the responses of the entire face and cannot obtain the useful intermediate information provided by the first and second hops. The feature responses from the first and second hops are fused in such a way as to capture the responses from different regions of the face, like eyes, nose, and mouth. All the responses of the third hop are used for classification. Binary classifiers are trained in each hop. Finally, the decision is combined and fed to a meta-classifier to predict the gender class. Logistic regression is used as a classifier.

Experiments have been conducted on the LFW and CMU Multi-PIE datasets, for which FaceHop achieves a classification accuracy of 94.63% and 95.12%, respectively. The model sizes for the respective models are only 16.9K and 17.6K, while LeNet-5 has a model size of 75.8K. Despite having fewer parameters, FaceHop still outperforms LeNet-5.

4.4.2 *DefakeHop*

Fake image and video content (known as "Deepfakes") is on the rise. Furthermore, the quality of the fake content has improved over time, making it more difficult to distinguish from real media. Similar to other tasks in computer vision, there are several traditional and deep learning methods for automatic Deepfake detection. DefakeHop [6] is an SSL-based method for fake face detection. It is lightweight in terms of model parameters, yet achieves state-of-the-art performance when evaluated on benchmark datasets.

The system diagram of DefakeHop is shown in Fig. 4.36. It consists of four steps: face image preprocessing, image feature extraction using PixelHop++, feature distillation, and ensemble classification. The preprocessing step consists of sampling face images from video, extracting face landmarks, face alignment, and cropping patches of size 32×32 from different parts of the face. The PixelHop++ method is then used to extract features from the local patches. The input to PixelHop++ is the 32×32 color images that focus on different parts of the human face. Three PixelHop++ units are used in cascade. To reduce the spatial redundancy, a (2×2)-to-(1×1) max pooling operation is performed between hops.

The next step of feature distillation consists of spatial dimension reduction and channel-wise soft classification. Principal component analysis (PCA) is applied on the spatial responses of every hop to reduce the spatial dimension prior to classification. Then, for every feature channel, a soft binary classifier is trained

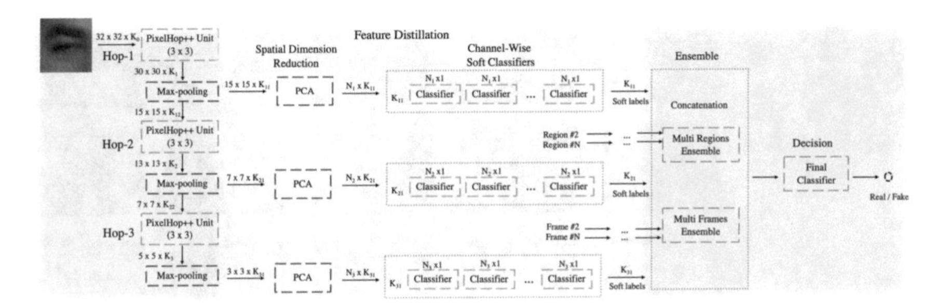

Fig. 4.36 System diagram of DefakeHop. Reproduced with permission [6]. Copyright © 2021, IEEE

that predicts the probability of that particular channel being fake. The ensemble classification step consists of a multi-region ensemble and multi-frame ensemble, followed by the final decision classifier. The classifier used is XGBoost. For the multi-region ensemble, the soft decisions of different facial regions are concatenated together, while in the multi-frame ensemble, temporal information is considered by concatenating the current frame with the three previous and three following frames. The probability of the whole video clip being fake is then obtained by averaging the probability of all frames.

The evaluation metric used is the area under ROC curve. DefakeHop achieves impressive area under ROC curve performance on benchmark datasets like UADFV (100%), Celeb-DF v1 (94.95%), and Celeb-DF v2 (90.56%).

4.4.3 AnomalyHop

Anomaly localization is the process of identifying anomalous regions within an image at the pixel level. It is of importance in applications like monitoring manufacturing processes, medical image diagnosis, video surveillance, and so on. AnomalyHop [41] proposes a method to generate anomaly maps from images. The training of AnomalyHop involves the PixelHop++ feature extractor. During training, only images that are free from anomalies are used. The system diagram of AnomalyHop is shown in Fig. 4.37. It consists of three modules: feature extraction using PixelHop++, modeling of normality feature distributions, and anomaly map generation.

First, the PixelHop++ method is used to extract patch-wise features from the images. The next step is to model the normality feature distributions. For this, three Gaussian models are used. The first is a location-aware Gaussian model. It models the distribution of the features of patches centered at similar locations across all the training images using a multivariate Gaussian distribution. The second distribution is a location-independent Gaussian model. Here, a single Gaussian model is trained for all local image features at each hop. This accounts for images of the same

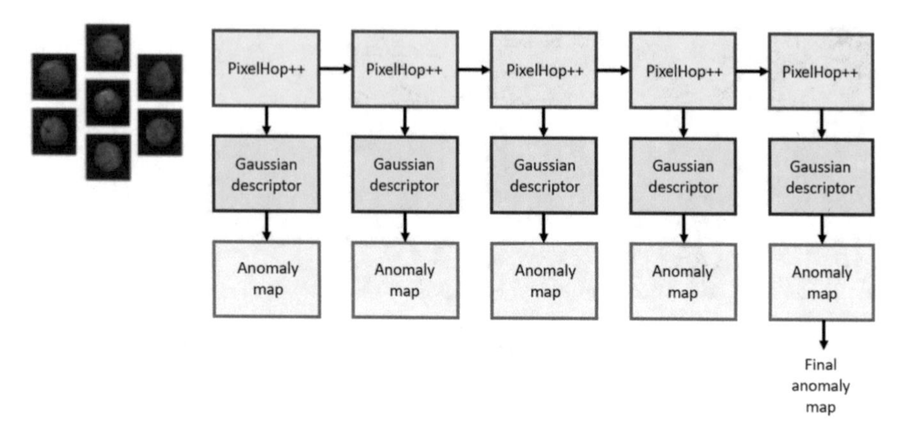

Fig. 4.37 AnomalyHop system diagram

texture class that have strong self-similarity and are shift-invariant. Images of the same class usually have intraclass variations, which are not captured by the first two distributions. This problem is addressed by the third distribution, which is called the self-reference Gaussian model. This Gaussian model is trained from the features of a single normal image. Finally, during testing, the Mahalanobis distance is used to measure the anomaly score of a patch. Anomaly scores are obtained for every location in a hop, thereby giving an anomaly map for every hop. The anomaly maps of all hops are rescaled to the same spatial size and fused by a weighted average method to generate the final anomaly map for the test image.

AnomalyHop is easy to train and is fast in the inference stage. It achieves an area under ROC curve of 95.9% on the MVTec AD dataset, which is the best performance among several benchmarking methods.

References

1. Achlioptas, P., Diamanti, O., Mitliagkas, I., Guibas, L.: Representation learning and adversarial generation of 3d point clouds **2**(3), 4 (2017). arXiv preprint arXiv:1707.02392
2. Aoki, Y., Goforth, H., Srivatsan, R.A., Lucey, S.: PointNetLK: Robust & efficient point cloud registration using PointNet. In: Proceedings of the IEEE/CVF Conference on Computer Vision and Pattern Recognition, pp. 7163–7172 (2019)
3. Besl, P.J., McKay, N.D.: Method for registration of 3-d shapes. In: Sensor Fusion IV: Control Paradigms and Data Structures, vol. 1611, pp. 586–606. International Society for Optics and Photonics (1992)
4. Breiman, L.: Random forests. Mach. Learn. **45**(1), 5–32 (2001)
5. Chang, A.X., Funkhouser, T., Guibas, L., Hanrahan, P., Huang, Q., Li, Z., Savarese, S., Savva, M., Song, S., Su, H., et al.: ShapeNet: an information-rich 3d model repository (2015). arXiv preprint arXiv:1512.03012
6. Chen, H.S., Rouhsedaghat, M., Ghani, H., Hu, S., You, S., Kuo, C.C.J.: DefakeHop: a light-weight high-performance Deepfake detector. In: 2021 IEEE International Conference on Multimedia and Expo (ICME), pp. 1–6. IEEE (2021)

7. Chen, Y., Kuo, C.C.J.: PixelHop: A successive subspace learning (SSL) method for object recognition. J. Vis. Commun. Image Represent. **70**, 102749 (2020)
8. Chen, Y., Rouhsedaghat, M., You, S., Rao, R., Kuo, C.C.J.: PixelHop++: a small successive-subspace-learning-based (SSL-based) model for image classification. In: 2020 IEEE International Conference on Image Processing (ICIP), pp. 3294–3298. IEEE (2020)
9. Chen, Y., Xu, Z., Cai, S., Lang, Y., Kuo, C.C.J.: A SaaK transform approach to efficient, scalable and robust handwritten digits recognition. In: 2018 Picture Coding Symposium (PCS), pp. 174–178. IEEE, Piscataway (2018)
10. Cortes, C., Vapnik, V.: Support-vector networks. Mach. Learn. **20**(3), 273–297 (1995)
11. Curless, B., Levoy, M.: A volumetric method for building complex models from range images. In: Proceedings of the 23rd Annual Conference on Computer Graphics and Interactive Techniques, pp. 303–312 (1996)
12. Eldar, Y., Lindenbaum, M., Porat, M., Zeevi, Y.Y.: The farthest point strategy for progressive image sampling. IEEE Trans. Image Process. **6**(9), 1305–1315 (1997)
13. Hassani, K., Haley, M.: Unsupervised multi-task feature learning on point clouds. In: Proceedings of the IEEE International Conference on Computer Vision, pp. 8160–8171 (2019)
14. Kadam, P., Zhang, M., Liu, S., Kuo, C.C.J.: Unsupervised point cloud registration via salient points analysis (SPA). In: 2020 IEEE International Conference on Visual Communications and Image Processing (VCIP), pp. 5–8. IEEE, Piscataway (2020)
15. Kadam, P., Zhang, M., Liu, S., Kuo, C.C.J.: R-PointHop: A green, accurate and unsupervised point cloud registration method (2021). arXiv preprint arXiv:2103.08129
16. Katsavounidis, I., Kuo, C.C.J., Zhang, Z.: A new initialization technique for generalized Lloyd iteration. IEEE Signal Process. Lett. **1**(10), 144–146 (1994)
17. Krishnamurthy, V., Levoy, M.: Fitting smooth surfaces to dense polygon meshes. In: Proceedings of the 23rd Annual Conference on Computer Graphics and Interactive Techniques, pp. 313–324 (1996)
18. Kuo, C.C.J.: Understanding convolutional neural networks with a mathematical model. J. Vis. Commun. Image Represent. **41**, 406–413 (2016)
19. Kuo, C.C.J., Chen, Y.: On data-driven SaaK transform. J. Vis. Commun. Image Represent. **50**, 237–246 (2018)
20. Kuo, C.C.J., Zhang, M., Li, S., Duan, J., Chen, Y.: Interpretable convolutional neural networks via feedforward design. J. Vis. Commun. Image Represent. **60**, 346–359 (2019)
21. Li, J., Chen, B.M., Hee Lee, G.: So-net: Self-organizing network for point cloud analysis. In: Proceedings of the IEEE Conference on Computer Vision and Pattern Recognition, pp. 9397–9406 (2018)
22. Li, Y., Bu, R., Sun, M., Wu, W., Di, X., Chen, B.: PointCNN: Convolution on x-transformed points. In: Advances in Neural Information Processing Systems, pp. 820–830 (2018)
23. Lin, R., Zhou, Z., You, S., Rao, R., Kuo, C.C.J.: From two-class linear discriminant analysis to interpretable multilayer perceptron design (2020). arXiv preprint arXiv:2009.04442
24. Moenning, C., Dodgson, N.A.: Fast marching farthest point sampling. Tech. rep., University of Cambridge, Computer Laboratory (2003)
25. Qi, C.R., Su, H., Mo, K., Guibas, L.J.: PointNet: Deep learning on point sets for 3d classification and segmentation. In: Proceedings of the IEEE Conference on Computer Vision and Pattern Recognition, pp. 652–660 (2017)
26. Qi, C.R., Yi, L., Su, H., Guibas, L.J.: PointNet++: Deep hierarchical feature learning on point sets in a metric space. In: Advances in Neural Information Processing Systems, pp. 5099–5108 (2017)
27. Rouhsedaghat, M., Monajatipoor, M., Azizi, Z., Kuo, C.C.J.: Successive subspace learning: an overview (2021). arXiv preprint arXiv:2103.00121
28. Rouhsedaghat, M., Wang, Y., Ge, X., Hu, S., You, S., Kuo, C.C.J.: FaceHop: a light-weight low-resolution face gender classification method (2020). arXiv preprint arXiv:2007.09510
29. Rouhsedaghat, M., Wang, Y., Ge, X., Hu, S., You, S., Kuo, C.C.J.: FaceHop: a light-weight low-resolution face gender classification method. In: International Conference on Pattern Recognition, pp. 169–183. Springer, Berlin (2021)

30. Schwartz, R., Dodge, J., Smith, N.A., Etzioni, O.: Green AI (2019). arXiv preprint arXiv:1907.10597
31. Turk, G., Levoy, M.: Zippered polygon meshes from range images. In: Proceedings of the 21st Annual Conference on Computer Graphics and Interactive Techniques, pp. 311–318 (1994)
32. Wagstaff, K., Cardie, C., Rogers, S., Schrödl, S., et al.: Constrained k-means clustering with background knowledge. In: ICML, vol. 1, pp. 577–584 (2001)
33. Wang, Y., Solomon, J.M.: Deep closest point: Learning representations for point cloud registration. In: Proceedings of the IEEE/CVF International Conference on Computer Vision, pp. 3523–3532 (2019)
34. Wang, Y., Solomon, J.M.: PRNeT: self-supervised learning for partial-to-partial registration (2019). arXiv preprint arXiv:1910.12240
35. Wang, Y., Sun, Y., Liu, Z., Sarma, S.E., Bronstein, M.M., Solomon, J.M.: Dynamic graph CNN for learning on point clouds (2018). arXiv preprint arXiv:1801.07829
36. Wold, S., Esbensen, K., Geladi, P.: Principal component analysis. Chemom. Intell. Lab. Syst. **2**(1–3), 37–52 (1987)
37. Wu, Z., Song, S., Khosla, A., Yu, F., Zhang, L., Tang, X., Xiao, J.: 3D ShapeNets: a deep representation for volumetric shapes. In: Proceedings of the IEEE Conference on Computer Vision and Pattern Recognition, pp. 1912–1920 (2015)
38. Yang, J., Li, H., Campbell, D., Jia, Y.: Go-ICP: A globally optimal solution to 3D ICP point-set registration. IEEE Trans. Pattern Anal. Mach. Intell. **38**(11), 2241–2254 (2015)
39. Yang, Y., Feng, C., Shen, Y., Tian, D.: FoldingNet: Point cloud auto-encoder via deep grid deformation. In: Proceedings of the IEEE Conference on Computer Vision and Pattern Recognition, pp. 206–215 (2018)
40. Yi, L., Kim, V.G., Ceylan, D., Shen, I., Yan, M., Su, H., Lu, C., Huang, Q., Sheffer, A., Guibas, L., et al.: A scalable active framework for region annotation in 3d shape collections. ACM Trans. Graph. **35**(6), 210 (2016)
41. Zhang, K., Wang, B., Wang, W., Sohrab, F., Gabbouj, M., Kuo, C.C.J.: AnomalyHop: an SSL-based image anomaly localization method (2021). arXiv preprint arXiv:2105.03797
42. Zhang, M., Kadam, P., Liu, S., Kuo, C.C.J.: Unsupervised feedforward feature (UFF) learning for point cloud classification and segmentation. In: 2020 IEEE International Conference on Visual Communications and Image Processing (VCIP), pp. 144–147. IEEE, Piscataway (2020)
43. Zhang, M., Wang, Y., Kadam, P., Liu, S., Kuo, C.C.J.: PointHop++: a lightweight learning model on point sets for 3d classification. In: 2020 IEEE International Conference on Image Processing (ICIP), pp. 3319–3323. IEEE, Piscataway (2020)
44. Zhang, M., You, H., Kadam, P., Liu, S., Kuo, C.C.J.: PointHop: An explainable machine learning method for point cloud classification. IEEE Trans. Multimedia **22**(7), 1744–1755 (2020)
45. Zhao, Y., Birdal, T., Deng, H., Tombari, F.: 3D point capsule networks. In: Proceedings of the IEEE Conference on Computer Vision and Pattern Recognition, pp. 1009–1018 (2019)
46. Zhou, Q.Y., Park, J., Koltun, V.: Fast global registration. In: European Conference on Computer Vision, pp. 766–782. Springer, Berlin (2016)

Chapter 5
Conclusion and Future Work

Abstract 3D point clouds serve as an important data format in intelligent systems. Researchers are continually developing new ways of making point cloud analysis more effective and efficient. It is common for new researchers to focus only on Deep learning methods while lacking a solid foundation of the fundamental knowledge of traditional methods. However, the traditional point cloud processing methods are the root of Deep learning methods, and they are still widely used in the industry. In this book, we complete a detailed analysis on point cloud processing, covering traditional methods, Deep learning methods, and our own explainable machine learning methods. In this chapter, we first summarize the book by discussing the advantages and disadvantages of the three types of methods. We hope the comparison and analysis of the three types of methods will help readers to gain a deeper understanding of this field. Next, we will provide some highlights for the future works in the field of point cloud learning, which may bring some insights to new researchers.

5.1 Conclusion

Traditional methods are based on local geometric properties of points and handcrafted features, where the latter belongs to the machine learning technique. Geometric reasoning methods or purely mathematical methods offer fast computation times and achieve good results in simple scenarios. However, they are sensitive to noise. Since the point clouds often have uneven densities and occlusions, this kind of methods cannot work well with complex scenes. Machine learning techniques rely on the results of handcrafted feature extraction processes, as we introduce through the feature detection and description tasks. Although machine learning techniques give better results than traditional methods, they are usually slower. In general, the traditional methods are interpretable, and they can be generalized to different tasks easily. Thus, they still have an important role in point cloud processing for real-world applications.

S. Liu et al., *3D Point Cloud Analysis*,
https://doi.org/10.1007/978-3-030-89180-0_5

Deep learning methods achieve outstanding performance but require massive amounts of training data (likely labeled data in a supervised learning setting). The key difference between Deep learning methods and traditional methods is that the former can perform end-to-end learning. Given a point cloud as the input, the output will be what we want, e.g., a label for each point in segmentation. The process is like a black box, which reduces the work on the human end. The existence of CNNs and MLPs and the optimization of parameters by gradient descent helps to realize this end-to-end purpose. With the help of GPUs, these methods can be very successful. However, the Deep learning methods are often criticized for big data, end-to-end, and GPUs. First, Deep learning relies too much on big data and supervision. Without big data and expensive label annotation, it is hard to say whether Deep learning methods would still obtain such good results. Second, the time and money cost are huge, so that it is very expensive to do Deep learning. Third, the end-to-end learning mechanism also limits the generalization ability of the method. Task-agnostic Deep learning methods are developing; however, most of the Deep learning methods are still task specific.

Explainable machine learning methods are based on the successive subspace learning (SSL) design principles. They offer a complementary approach to conventional deep-learning-based design. These methods are designed with consideration to the advantages and disadvantages of both traditional and Deep learning methods. First, this set of methods is data-driven but not data eager. More data helps explainable machine learning methods, but they are also robust with less data. Second, the feature extraction process is unsupervised, and no backpropagation of parameters is needed. Hence, the training is quite fast and cheap compared with Deep learning. Finally, the whole process is interpretable or explainable, and the extracted features can be used in multiple scenarios. For example, our proposed feature learner can be used in object classification, part segmentation, and registration tasks.

In conclusion, the three types of methods each have their own advantages and shortcomings. Some are more suitable for a specific purpose than others, and they should be used in this way. We hope this book will provide new researchers with a deeper understanding of the point cloud processing techniques.

5.2 Future Work

For researchers who are going to explore this field in more detail, one possible direction that is not covered in this book is large-scale point cloud learning. All three methods are currently limited to small-scale point clouds. Traditional methods have poor efficiency for large-scale point clouds. Most of the Deep learning methods split the large-scale point clouds into smaller pieces to reduce memory consumption. Nevertheless, more researches related to directly processing large-scale point cloud has emerged recently. The explainable machine learning methods can easily be extended to large-scale point cloud learning, because they do not rely on GPUs for processing. We are extending the explainable machine learning methods to solve tasks like point cloud odometry and 3D object detection in urban scenes and semantic segmentation in indoor scenes.

Index

S. Liu et al., *3D Point Cloud Analysis*,
https://doi.org/10.1007/978-3-030-89180-0